HOMO SAPIENS DIVINE

Life's quest for perfection: From a strand
of protein to the exalted human and beyond

HOMO SAPIENS DIVINE

Life's quest for perfection: From a strand
of protein to the exalted human and beyond

Varghese Mani

Notion Press

Old No. 38, New No. 6
McNichols Road, Chetpet
Chennai - 600 031

First Published by Notion Press 2016
Copyright © Varghese Mani 2016
All Rights Reserved.

ISBN 978-1-945621-25-3

This book has been published with all efforts taken to make the material error-free after the consent of the author. However, the author and the publisher do not assume and hereby disclaim any liability to any party for any loss, damage, or disruption caused by errors or omissions, whether such errors or omissions result from negligence, accident, or any other cause.

No part of this book may be used, reproduced in any manner whatsoever without written permission from the author, except in the case of brief quotations embodied in critical articles and reviews.

Dedication

*To the memory of my parents,
Mr M.V. Mani and Mrs Achamma Mani*

Homo Sapiens Divine

Life's quest for perfection: From a strand of protein to the exalted human and beyond

Prof. Dr. Varghese Mani

"Man may be excused for feeling some pride at having risen, though not through his own exertions, to the very summit of the organic scale; and the fact of his having thus risen, instead of having aboriginally placed there, may give him hopes for a still higher destiny in the distant future."

The Descent of Man by Charles Darwin

Contents

Foreword   *xiii*

Acknowledgements   *xv*

Prologue   *xvii*

Introduction   *xxxi*

1. Matter Wriggles and Life Emerges   1
2. Life Tickles/Perception   19
3. Cognition and Confusion   33
4. Conscience and Consciousness   65
5. The Ladder/Tree of Evolution   89
6. Lucy to Darwin   131
7. Beauty, a Lure   163
8. Homo Sapiens Divine   183
9. Culture and Existence   221

Epilogue   *235*

Bibliography   *237*

Index   *257*

Foreword

I was looking forward to Professor Varghese Mani's opus magnum "Homo sapiens divine" – Life's quest for perfection, for I knew it was coming. I was honoured when I was asked to write a foreword to this remarkable book, so different from anything that I have read before. Like the Greek God Janus it has the courage to look into the future, the evolution of man and his progress into the unforeseeable future.

Mortal man's quest for perfection and immortality has been debated by philosophers and denigrated by scientists since time immemorial. The classic example is the Legend of the Flying Dutchman, who was cursed to immortality until he could find a woman who loved him enough to die with him. In the same vein there is the Greek legend of Tithonus and Eos. Tithonus, a mere mortal, fell in love with Eos, the goddess of love, she asked Zeus to grant Tithonus immortality. Zeus being jealous of this attachment, granted him immortality but without eternal youth, and so the relationship between the two lovers soured.

The other side of this coin, which this work reveals, is the possibility of the explosion of scientific knowledge, genetic engineering and what may come in the future if things are left to human ingenuity and imagination. The author has touched on various and diverse aspects of man's journey through the evolutionary process in a systematic manner, there are parts, which belong to the realms of fantasy. The book is informative and thought provoking. It is both philosophical and scientific. It is not a book meant for casual bedside reading. However, it is certainly a book that I would like to have on my shelf, to seriously browse into from time to time and to refer to when the occasion arises

H.S. Adenwalla
Emeritus Professor of Surgery
Head of the department of Plastic Surgery, Burns and The Charles Pinto Centre for Cleft Lip, Palate and Craniofacial Anomalies.
Member, Smile Train Medical Advisory Board (South Asia and New York).

Acknowledgements

I am honored that Dr H.S. Adenwalla has written the foreword for this book. He is a great surgeon, philosopher and philanthropist. He has devoted his life for children suffering from cleft deformities and syndromes of the craniofacial region. No words will suffice to express my gratitude to him.

My colleagues, friends and students, being aware of my interest in evolution, genetics, anthropology and facial beauty were the ones who persuaded me to write this book. I am grateful to them for their friendly nudges, comments and suggestions.

I am indebted to my teachers, mentors, and great authors and scientists from whom I have drawn inspiration and learned. Apart from my work on the evolution of beauty of human face, the content is based on the research findings of scientists and opinions of philosophers.

Notion Press has kindly consented to publish this book. The publishing team has done a tremendous job in editing this book and making it devoid of syntactical errors. They have done the cover design as if they have read my mind. Notion press team have done an excellent job. I am immensely thankful to them.

My wife, Mini, a biology graduate, despite being a strong theist, tolerated my agnostic stance and helped me by giving suggestions and critical comments while preparing this book. My children, Dipu, Alias, Minu and Kanchana, went through the manuscript, gave me creative suggestions and were a great moral support. My grandchildren, Niya, Nora, Liza, Basil and Leya, were a welcome relaxation during difficult times. If not for the support of my family, this book would not have been completed.

Prologue

Biological Robot

Nimi was born in the year 2050. Now, in 2550, she was 500 years old. When she was thirty, the world was raging with a controversy. Back then, she was a specialist in biotechnology—a field that had progressed to the extent that life could be prolonged and youthfulness maintained. Many elders and some philosophers had argued vehemently against the implementation of this technology in humans, and warned that this could create havoc in society—existential problems, class differences and even extinction of the human race. Their warnings and protests were in vain—they grew old and became extinct. The world was then taken over by the technologists. Eugenics was widely practiced. Within a short period of time, a new world order was instituted.

Nimi, a distinguished biotechnologist, was a proponent of eugenics as she saw immense possibility in it. The human genome was completely decoded in the year 2003. Genetic engineering has progressed to the degree that life was 'produced' in the laboratory. By creating artificial genes, bacteria and other living forms many chemicals and proteins could be manufactured. By this time, it was but commonplace for artificial meat and vegetables to be manufactured in laboratories.

Gene therapy could make humans live longer. Cures for almost all maladies, including cancer, were in place. Biotechnology apart, all other sciences were also progressing at a rapid pace. Colonization of other planets has started, with humans going there initially as tourists. Interplanetary travel was routine. Problems such as climate changes due to global warming, and excessive radiation due to loss of the ozone layers, and so on had created some mayhem, so all living quarters and public places were now regulated with oxygen supply and climate conditioning.

The system of marriage had disintegrated, and free sex was the norm. However, no one was allowed to have children. Depending on the society's requirement, children were produced in the labs, with genomes ranging from pageboys to rulers printed into them. They were then reared in nurseries.

When Nimi was thirty-five years old, a new technology name 'brain rewiring' was introduced. Life was already extended for an indefinite period; this technology created the possibility of changing a person's capabilities completely. By rewiring the brain, all the information in the brain was deleted and new brain connections were established with new knowledge. The person forgot everything about his past—it was like a rebirth. A new person with totally different capabilities was created. This technology was initially available only for the rich and powerful, but as social systems changed and technology advanced, it was available for general public as well.

Nimi was finding it hard to cope with the fast-developing technologies. So, she decided to get rewired—as a biotechnologist again. A person could take up a new profession with the latest knowledge built in to the brain. She lost all her friends and acquaintances when she became a totally new person.

The world order was drastically changing. Eugenics was able to produce human beings with larger brains with greater intelligence, and more senses such as that of bats (eco-location), dogs (enhanced smell and decibel range), eels (electrical discharges) and migratory birds (navigation).

With the profusion of such changes, people who had smaller brains and limited capabilities could not keep up. So, the option of getting rewired to a profession of their choice was restricted. A person's choice of profession depended on the number of neurons in his/her brain, which was roughly proportional to the brain size. It was mandatory for a person to get rewired periodically if he wanted to get the life-extending therapy. Most of the human beings produced by eugenics had larger brains with more neurons, and they take on higher-ranking, powerful jobs. Important professions were allotted to artificially created human beings with minimal emotions, whose average brain size was as large as 2,000 cubic centimeters (cc) or more.

There was no need for humans to be engaged in manual labor as robots did most of the work—even the work of physicians and surgeons was done by robots. The removal and replacement of diseased parts were part of the usual dictum; organs could be reared from pluri-potential or even somatic cells of the individual who sought transplantation.

Nimi had to change her profession several times through rewiring. So, Nimi, who held high posts once upon a time because of her talent, education, hard work and research acumen, has climbed down the ladder. About ten years ago, she was rewired as a poet.

Poets were relegated to the lower rungs, as they were not very useful for the highly technological society. They were considered as being highly emotional and sensitive, and were seen as living in a world of dreams. While rewiring a person to make him/her a poet, a part of the person's brain was left as such, so that he/she could have free thought, dreams and imagination. This was a topic of debate among policy makers.

For poets, some portions of the brain were left as such during the rewiring—especially amygdala and hippocampus, which dealt with emotions. This was the problem that affected Nimi—she was an emotional person, and emotions were considered as weaknesses.

It was time for Nimi to get her brain rewired. She had the option to continue as a poet or take up another menial job. Either way, she had to either get rewired or move to the fringes and grow old and die a natural death. She introspected on her choices. 'Fringes' was the area demarcated for those who decided not to get rewired. They could not get the life-extending therapy, but they were looked after well, and doctors regularly visited them.

Nimi was a bright, intelligent and attractive girl who had contributed substantially to the field of biotechnology, and had held high posts. However, she was now in the lower rung of society, and her contributions were known only in the annals of history. Though she had been regarded as 'beautiful' during her 'earlier life', she was now considered unattractive, going by the changed norms of beauty (large head, small jaws and exaggerated oval face).

She was frustrated, as people did not give her any importance. Some looked at her with pity, and some ridiculed her. When her frustration reached its peak, she wrote poetry and posted it on the sites. Very few people viewed her poems—mainly other poets and writers. Once every three months, the writers' guild met and discussed, criticized, appreciated and shared their frustrations. They lamented society's deteriorating values. They were concerned about their fate. Many of them were thinking about not getting rewired, for every rewiring often meant going down to a lower segment.

After one such meeting, Nimi wondered about what she should do. The policy was to save the previous brain wiring records in compact chips and hand over a copy to the person. He/she could then put them into a chip reader, put the reader in the specialized helmet and then view their previous life as if it was a movie. She took out the chips in which her previous lives were coded. She took the oldest one and loaded it into the chip reader, placed the reader over her scalp and switched it on. She viewed her first life as if it were a dream. The first four years were hazy. She fast-forwarded it and viewed the important scenes until she was thirty-five years old—when she had decided to get rewired.

Nimi's First Life

Nimi's mother was a professor of history. Her father was a reputed and affluent businessman. Nimi lived a disciplined life. She always topped her class in studies. She went to become a reputed biotechnologist, and later the head of an internationally reputed research center that contributed to the progress of biotechnology.

Nimi's parents valued ethics. Her mother wrote profusely against eugenics and argued that interfering with natural evolution would ultimately create havoc, and that it could even lead to the extinction of the human species. Though nobody believed in a supreme creator anymore, there was a belief, which many elders held, that the entire universe was a living system.

Nimi was a vehement supporter of eugenics. Several arguments broke out between mother and daughter due to this, but both of them were careful that the argument didn't affect them on a personal level. Her father, by nature or by intent, passed some casual comments to reduce the tension.

Nimi was literally wedded to her profession, and she remained single. She had an important role to play in the development of eugenics—her research in adding genes to the chromosomes, and getting its expression as new characteristics especially in higher mammals was highly acclaimed. This paved the way for the possibility of enhancing the brain's capacity and functions. Scientists were also able to add many more senses that could be shut and opened as per the requirement. Eugenics could increase or decrease the number of chromosomes, add genes, and extend life indefinitely. Genes and chromosomes became playthings in the hands of the biotechnologists, who were soon capable of creating bizarre organisms. However, the law has prevented this, as the monsters created could be disastrous for the whole world. The success of genetic engineering had many adverse effects—for one, the newly created humans had larger brains and immense power, and they took control of the world. The ones with smaller brains had to take up menial jobs—a pattern very similar to the racial discrimination of olden times.

Nimi's close friend, Dr. Emma, was a neuropsychologist. She was a reputed scientist who had worked with biotechnologists and scientists from other fields. She had conducted extensive research on neuro-biochemistry and the wiring of the neurons, and was well known in the field of neuroscience. She was part of the team that had succeeded in rewiring the brain. Initially, rewiring was performed on criminals and people who had mental disorders. The possibilities were immense and far-reaching. Within a short period of time, rewiring of the brain to enhance the brain's functions and store more information and knowledge was started. Emma and Nimi were part of the first group of top professionals who underwent rewiring of the brain. During the first rewiring, their memory was not completely erased. However, in the following years, procedures were changed as the new generation of artificially reared humans took over the reins. Rewiring now meant a complete removal of all old memories, with the brain being filled with new information. Rewired humans thus became 'biological robots'—they had no emotions and no friends; they only mingled with professional acquaintances.

Nimi went through her past life—her childhood and adolescence. She viewed the attachment she had with her parents and friends during her days in school and university, and her love affair with a colleague—the emotion of love, which was now alien to her. When her boyfriend had died in an accident, she had found it very hard to contain her grief. In spite of her rewired brain, she slowly began to experience the emotions. She felt grateful for her position as a poet because it had left some space in her brain for emotions. However, the brain is an enigma—whatever rewiring may be done, the brain has a tendency, due to its nature, to make fresh connections. So, when Nimi was viewing her previous life, new connections were being established in her memory.

Though 500 years had passed, Nimi was physically only thirty-five years old. After a long time, she felt like crying. She had heard crying only in the eugenic nursery, where infants were reared. At first, she felt a bit ashamed but crying made her feel as if she had been released from a cage. She cried continuously, as if she were washing off dirt from her inner self by doing so.

It was time for Nimi to get rewired—and she had to go to a lower rung in the social ladder. Emma had decided to quit about five years back. Because of the rewiring, both of them had forgotten about their old friendship. Nimi now recollected her friendship with Emma, and searched for her in cyberspace. She found out that Emma was living in the fringes.

Nimi had to take a decision about rewiring within ten days. The reminder alarm chimed in the chip, which was implanted inside her scalp: "Ten days to rewire or quit." She decided to go to the fringes and meet Emma.

She had to get permission to do so. She wired for permission. All the documents were copied in the center and permission was granted in no time. The authorities actually encouraged people who decided to quit. She got out of the apartment and climbed into her aero-car, programmed it and pushed the 'Go' button. In about five minutes, she reached the fringes.

The 'fringes' were at the boundary of so-called civilized life. The scene was like that of an old-time village. A few old people were taking an evening stroll among the trees—such greenery was a rare sight in the civilized world. Nimi remembered her good old childhood. Emotions welled up within her, and her heart jumped with an unknown joy. Emma had already been alerted from the center, and she was waiting for Nimi. There was a beautiful garden in front of Emma's quarters. The garden was full of flowers and butterflies fluttered all around. The music of the birds was exhilarating. The cool breeze brought in fragrance and a divine solace. Nimi's eyes welled up with tears as she inhaled deeply.

Emma spread her arms and Nimi fell into them as if she were a child, and started weeping. Emma didn't say anything; she just patted Nimi's back to console her. It was quite long since Nimi had seen any old people, and the thought of getting old was scary.

"Emma, how did you recognize me?"

"I read some of the old chips."

Both of them laughed.

Emma said, "You write poetry, and I write prose, novels and short stories. In between my activities,

I used to go through my past life. It was gradually causing some kind of rewiring in me. So, I have even started writing some critical pieces against rewiring and eugenics. Though we vehemently supported eugenics at one time, we forgot that emotions and consciousness have developed as natural selections, and that they have survival benefits. Nature is all-powerful. If this goes on, a day may come when the human race will be eliminated from the face of the planet. Biological robots cannot exist for long by challenging nature. Don't you know that there were boom and bust cycles on Earth? We have to remember that a vast majority of the species that once existed became extinct due to the natural calamities. Why should humans be exempt?"

Nimi was not convinced. She could not perceive that this would happen to the human race. She said, "Emma, not all the species were wiped out during the extinction events. These events affected certain areas more than others. We have to realize that humans are a global species—maybe the only species that is global. Domesticated animals are also global, but they are global because of humans. So, an extinction event may not wipe out the human race. However, I fully agree that it will be a big blow to the species, and we may find it very hard to recover from it. Yet, by then, some humans may have migrated to other planets, and even interstellar travel might have become possible. In the way that they became global by migrating from Africa to the rest of the planet, they may migrate to other parts of the universe and even become a universal species."

Emma said, "Yes, but humans are very fragile creatures now because a majority of the ones out there were artificially created. They did not evolve through natural selection. They are all living in an artificially created environment, so they are incapable of resisting any natural calamity. As you said, they may have to find new abodes. I heard that they are planning to have only sexless humans—in other words, humans will only be produced in the labs. Sex was a very important step in the process of evolution. I feel Freud was right in postulating that sex was the most important driving force, because of the innate libido of living beings to propagate."

Nimi said, "If the KT event had not taken place, the dinosaurs would have continued as the dominant species, and mammals would not have had a chance to become dominant. Birds, in spite of their small brains, are very intelligent. They evolved from the dinosaurs, and they are the only group, except for bats and insects, that have conquered the sky by natural process. I have a strong feeling that humans are an accidental species."

Emma was silent for a while. Then, she said, "Nimi, you are right. A normal brain cell has around a thousand connections. If it were 10,000 or more, there would be no need for such huge investment in the brain. There would be no need for such a huge brain as well. We all know that brain size is not the only criterion for intelligence. I still cannot comprehend why intelligence has not developed in other vertebrates and non-vertebrates, coelenterates and so on to the degree it has in humans. Out there in the universe, there may be millions of varied organic as well as inorganic forms of life, with diverse senses and faculties that are beyond our imagination. We may be able to explain evolution, but the question of 'why' looms large. Nobody is asking this metaphysical question now—the existential dilemma does not bother anyone now."

She continued, "So, I tell you, Nimi, I am really happy here—I do a bit of gardening, help the older people, sing, dance, read and gossip. Do you know, Nimi, I have a collection of your poetry. I am leading a full life. The only regret I have is that there are no children here." Nimi listened keenly to the discourse, and realized that Emma was dousing most of her apprehensions.

Emma continued, "I am growing old without any regrets. Thanks to medical advances, it is possible to live without pain. By rewiring the brain, we became new individuals and forget our past. It is like a rebirth—similar to the thought prevalent in ancient India. However, in that hypothesis, if an organism had lived a good life, it was reborn as a superior being. Here, you are reborn as an inferior being and go down in the ladder."

After a short pause, Emma continued, "A world devoid of death is artificial. The present humans are all artificial—that is why I call them 'biological robots.' Every action is mechanical and logical, without any complexity or duality. Death and birth are the two boons humans were given. If there is no death or birth, there is no progress. There is progress in each generation—there is betterment of the brain by natural selection to the tune of nature. If you challenge it, you are doomed. We have not been here for even half a million years."

Nimi was silent. She could not help but agree with what Emma was saying. She was wondering to herself, "Is it possible to save this world from disaster?"

Emma continued, "Do you remember our first life, Nimi? We always had arguments and doubts about values and changing values. We lied, cheated, loved, hated, argued and played. We were sometimes confused, and therefore all the more emotional. When rewiring was perfected, we lost our emotions. We never lied, loved or hated. All these emotions were considered weaknesses and the emerging humans were not supposed to be weak.

"I consider myself very lucky because I opted to become a writer, and you are lucky that you opted to become a poet. The artists, though looked down on by the scientists, were bestowed with imagination and emotions during rewiring as these qualities were essential for poets and writers. Do you know, Nimi, that the new generation lacks intuition?"

Nimi asked, "Is there a way for us to save this world from the impending disaster—a world that cannot reproduce or love? You were a neuropsychologist and I was a biotechnologist—can we do something?"

Emma fell silent. Both of them contemplated what seemed to be an impossible proposition.

There was a short thud at the door. Emma's face brightened. She called out, "Come on, Inam. Meet my friend, Nimi. She is going to be your friend as well."

A chimpanzee came forward and grinned, and extended his right hand towards Nimi. She held it firmly and shook it.

Emma said, "He is a good friend. I am teaching him. I am putting the chips into him and he is learning fast.

It is a slow rewiring. He will become a great teacher in his community. One day, they will become an intelligent race by natural selection and take over the reins of the globe after nature wipes out the biological robots or they become extinct."

Introduction

Not only children but adults also wonder about the twinkling stars, the vivacious colored birds, flowers, butterflies and the harmony of nature. The vast expanse lingers in our minds as a riddle.

I doubt this riddle can be solved in the perceivable future, though man has achieved the capabilities to think, perceive, predict and act, and has gained beauty, awareness, intelligence and intuition through the evolutionary process.

Being a maxillofacial surgeon who is more into cosmetic surgery of the face, beauty and the evolution of beauty was a subject of fascination. Basic training in biology and a passion for biological evolution pushed me to study the changes that have taken place in the human face during the evolutionary process. The statement, 'Brain case is expanding at the expense of the jaws', points to the fact that the cranium is enlarging and the jaws are reducing in size during the evolutionary process.

I have used a modification of the cephalometric analysis (measurements of the skull and jaws) relating the cranium and face and applied it to the widely accepted beauty proportion—1:1.618, also known as 'golden' or 'divine' proportion. The proportion between the cranium and face slowly became approximated to the golden proportion as we evolved from Australopithecus afarensis (Lucy, the mother of all humans) to Homo sapiens, the modern man. By projecting the graph of biological evolution into the future, we can see that humans will reach near perfection in beauty and function in a million years. Over the evolutionary process from Australopithecus, a new species was named almost at every half-a-million-year mark. This means that two more Homo species may develop in another million years, if natural evolution is allowed to progress unhampered. We could name them 'Homo sapiens super', after Nietzsche's 'super man' and 'Homo sapiens divine' after the divine proportion, respectively—with more lofty faculties, which we cannot envisage now!

Beauty, an elusive property, is subjective and defies definition. Whether it depends on any other factors like sexual selection, survival and functional benefits, or whether it is a trail towards perfection and so on are to be pondered. A person who is healthy, vital, active and athletic is considered attractive. Attractiveness is almost akin to beauty. Certain other factors that, for sure, attract both males and females are features that have survival benefits. For example, large breasts are considered beautiful. It is a well-known fact that many women underwent silicone implants to augment their breasts to look more attractive (its popularity dropped after it was proved to be harmful). Large breasts give the impression that they can produce more milk, which is important for feeding the offspring. Likewise, large and wide buttocks send out two messages—the wider one is more attractive than the massive ones, and gives the impression that it can hold the large head of the foetus. A large head is an indication of greater intelligence. A large and bulging bottom means that it holds lot of fat, which can be utilized during lean periods

(Example: Steatopygia among certain Africa tribes). Large breasts and wide hips, which give an hour-glass appearance, are seen as attractive and beautiful features on women. When compared to the rest of the mammals, females are more attractive than males in humans. No wonder, the onus of selection of a partner is shifted from the females (in the lower forms of mammals) to males (in humans) to a considerable degree.

Females are more attracted to men who have athletic bodies rather than those who are body builders. Aggressive features usually deter females, though this may be helpful in intra-sex competitions. There is a general notion that females prefer men with deeper pockets as, in this competitive world of humans, material possessions are an indicator of power. Have we adjusted our concept of beauty accordingly?

From the time of evolution of mammals, the bits of information stored in the brain surpassed the information stored in the chromosomes. As man started speaking and writing, extracorporeal information excelled all others and continues to grow in geometrical ratio. Technological advancements are mind-boggling. Eugenics and other such biological manipulations may surpass natural evolution. Artificial 'biological robots' may rule the planet, and even the universe beyond, in the not so distant future.

This book is mainly intended to examine the progression of life from a strand of protein to the highly exalted human being and beyond.

Origin of matter and universe, origin of life and origin of consciousness and awareness are some mysteries that defy logic and comprehension. Definitive answers are yet to be found, and science can only postulate. Religions find solace from these ambiguities in God—'the Great Architect of the Universe', who is said to be omnipresent, omnipotent and omniscient.

Regarding the origin of the universe, the most accepted theory is the 'Big Bang' from a gravitational singularity that had infinite density, mass and space-time. 'What is singularity and how it was formed?' is a question nobody is able to answer conclusively. The question, 'What was there before singularity?' also remains unanswered. (This question itself was considered illogical, as there was no time before the Big Bang.) Just why the singularity burst into an expanding universe is also an unsolved riddle.

Initially after the Big Bang, there was only hydrogen and helium, the simplest of elements. Hot stars might have produced heavier elements like carbon, iron and oxygen, which were spewed into space by supernova explosions that produced material for a new generation of stars. There are more than a hundred billion stars in the Milky Way, our galaxy. A substantial fraction of these stars have planets. So, it is possible for many planets to have favorable conditions for the existence of life. Our solar system might have originated from the matter thrown out by supernova explosions of certain earlier stars about four-and-a-half billion years ago—that is, about ten billion years after the Big Bang. Earth was formed mainly by heavier elements, including carbon and oxygen, which were essential for the creation of life.

After ten billion years, the sun will swell up and engulf our planet, thereby destroying every form of life.

By then, intelligent forms of life on the planet should have mastered space travel and inhabited another favorable planet—if not, our form of life is doomed. We are able to create the building blocks of life in the laboratory from non-living materials, but we are unable to make ribonucleic acid (RNA) or deoxyribonucleic acid (DNA) capable of reproduction, till now.

The Big Bang model proposes that, in the beginning, all matter in the universe was contained in a single point known as 'singularity.' The Big Bang theory is based on general relativity theory. However, there were certain inaccuracies in the Big Bang theory due to its sole dependence on general relativity theory, and the 'String' theory was proposed. This theory put forth that two branes (membranes) much larger than the universe collided, thereby creating mass and energy. It is generally accepted that the universe started around fifteen billion years ago and that it may end after twenty billion years from now.

Correspondingly, the origin of life has not been conclusively explained. Many theories and explanations abound, and it is considered to be a chance occurrence in the primordial environment. Till now, no scientific method has been able to 'create' life. A similar environment was artificially created and experiments were conducted but life was not created, even though science was able to produce the building blocks of life artificially.

Awareness and intelligence are part of yet another riddle. Though some opine that only humans were endowed with these enigmatic properties, rudimentary forms of awareness were present in many mammals, and, during evolution, it was 'advanced' to 'better quality.' The question still remains: Why? Awareness created more stress to the animal—not a very positive factor for the quality of life though it may have certain survival benefits. Evolution of awareness is a mystery—evolution itself is an enigma.

Paleo-biological findings provide indubitable evidence for evolution from a strand of protein to the present complex, intelligent and intuitive human being. Over three-and-a-half billion years, life has progressed by trial and error, and rejection and selection. During this progression, movements have changed from chemotactic to conscious, activities from instinctive to intentional, and decisions from logically linear to intuitively complex. From passive existence emerged awareness and conscience, which may have survival benefits.

However, if a meteorite, like the one that hit the planet about sixty-five to seventy million years ago (the KT event, which caused the extinction of the dinosaurs, the dominant creatures on the planet), does hit the planet, it is doubtful whether humans will survive the impact. Smaller animals and lesser living forms like insects and bacteria may survive—they have survived greater adversities after all. Hence, development of intelligence cannot lay to claim great survival benefits, though it might have been helpful in competition with other predatory and physically powerful animals. Advancement of intelligence, awareness and intuition may have other purposes.

We can observe that, as we progressed through evolutionary processes, physical evolution has given way to intellectual evolution. Intelligence depends on the number of neurons in the neo-cortex of the brain, which is almost directly proportional to the brain size. The amount of information stored in the brain is much more than what is stored in the genome (the chromosomes of the organism) in higher animals like mammals while it is the opposite in lower forms like fish and amphibians.

For the early life forms, 'living' in the initial stages life was passive existence—getting adjusted, or adapting to the surrounding environment. Later, they evolved to create an environment suitable for themselves, and even started changing the environment for their needs. Analysis of evolution shows that the process is geared towards better function and beauty. Perfection is the point where beauty and function meet.

The only thing that survives after an organism is its 'genes', provided the organism had transferred its genes while it was alive. The progeny gets only fifty percent of the chromosome in the cell (gamete); the rest comes from the partner, which means that atman is akin to the gamete (the chromosome in normal cell is in pairs—one set each from both the parents). By a process called reduction division, the pairs separate and form two gametes, each having only one set). In the Indian concept of 'Arthanareeswaran' (half-woman and half-man), both sexes are essential for the completion of creation. If so, the offspring could be considered the physical rebirth of both the sexes. So, was the genome depicted as the 'soul' by sages in the past? If a person/ animal leads a good life—meaning, he has good genes, in which environment also plays a role—he will take birth as a superior being. This is exactly what happens in evolution. One who has superior characteristics/genes (from either one or both parents, or by favorable mutations) will be selected by nature in the struggle for existence. If we project it to a longer period (many generations), a totally different superior being (a new species) emerges when good characteristics are accumulated. This phenomenon is called 'evolution.'

A comparable parellel of evolution is seen in the ten incarnations of Lord Vishnu who sustains the universe, according to Indian mythology. The first was in the form of an aquatic fish. Science also agrees that life originated in water. The second incarnation was the tortoise—an amphibian. During the evolutionary process, life migrated from water to land, and developed the capacity to live in both water and on land. The third incarnation was as a pig—a lowly mammal. The fourth was 'Narasimha', which means half-man and half-lion. This could be considered as a portrayal of the transition of the lowly mammal to a higher mammal—the primate. The fifth was 'Vamana', a dwarf, and may be portraying the bush man. The sixth one was 'Parasu Rama'—he has an axe (parasu) as his tool and he is a woodcutter. The next was 'Sree Rama', who had a bow and arrow as his tool. This is evolutionary progress from a woodcutter to a hunter. The arms and tools are considered extra-corporial limbs. The reach of the second Rama was up to the tree top, which is farther than an axe. The third Rama, the eighth incarnation, was more noteworthy. He was 'Bala Rama' and his tool was a plough. A plough is meant to till the land. This depicts the migration from the forest to the plains as an agriculturalist. The next was 'Krishna' who had a 'chakra' (revolving disc) on his arm, which can be compared to a cruise missile that can follow the target and destroy it. Moreover he was a smart politician, with convincing arguments that would favor him and his associates. The tenth incarnation, 'Kalki', is yet to come, and is expected in this eon (Kali Yuga). He is said to come for the final judgment—to end the world, and start another cycle.

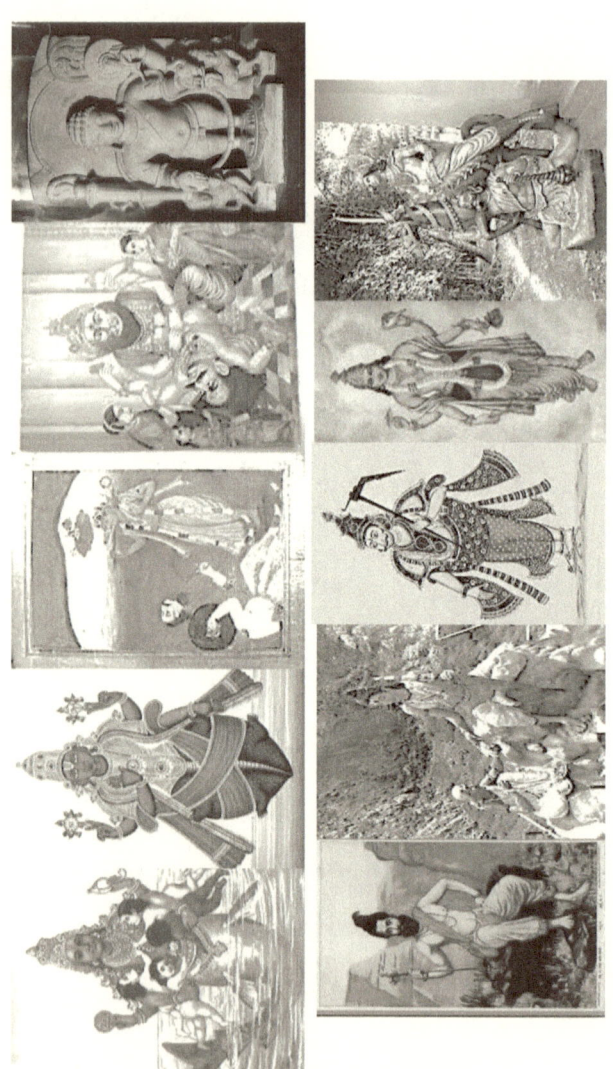

Fig: 1 The ten incarnations of Vishnu: Matsya (fish), Kurma (tortoise—amphibian), Varaha (pig—mammal), Narasimha (half-man and half-lion), Vamana (dwarf man), Parsu Rama (Rama with an axe), Sree Rama (Rama with a bow and arrow), Bala Rama (Rama with a plough), Krishna (Krishna with a flute, and a revolving disc as his armament), Kalki (the last of the ten incarnations).

1
Matter Wriggles and Life Emerges

"One can define Life to be an ordered system that can sustain itself against the tendency to disorder, and can reproduce itself. That is, it can make similar, but independent, ordered systems."

— **Stephen W. Hawking**

The pertinent question is: 'Why the universe and life?'

'How' and 'what' is related to scientific enquiry, but 'why' is primarily a metaphysical question and can be answered clearly only after 'what' and 'how' are answered. However, the metaphysical question can also be pursued simultaneously. That is what many philosophies and religions do. How life started and awareness and intuition emerged are questions that are yet to be answered conclusively. Theories abound and ambiguity still prevails. The search is on.

A self-sustaining dynamic nature is the most important characteristic of life. It undergoes metabolism; it grows and reproduces. Living organisms have the unique characteristic of reaction to a stimulus, and they adapt to their surroundings and gradually evolve to become better organisms.

Reproduction of the DNA occasionally had errors, which are called 'mutations.' Most of the mutations were deleterious, but a few of them were favorable for existence and these were selected. They provide survival benefits. This is the basic concept of the 'neo-Darwinian' principle of evolution. Initially, evolution was a slow process. About two-and-a-half billion years were required to evolve multicellular organisms from unicellular organisms, and another billion years for the evolution of mammals through fishes, amphibians and reptiles. Increase of evolutionary speed can be perceived from this, as well as the evolution of humans from mammals in a short period of just a hundred million years. Sexual reproduction could be one of the facilitating factors. There is continuation of genes in the evolutionary progress. This is evident from the embryological development of the organisms. Humans have the genes of the fishes and early mammals. Fishes have all the important organs of humans, and mammals have almost all of them. From early mammals to humans, it was more of a fine-tuning. Development of language, speech and writing has revolutionized the passing of information. The number of nucleic acids in our genome is about three billion—a good number of them are superfluous or inactive.

Theories of Creation

It is interesting to go through different prevalent concepts about the origin of life. Australian aborigines believed that the world originated during 'Dream Time', and that humans were created from shapeless bits of nature by astral beings that wandered the earth.

Judo-Christian

One of the most popular and prevalent theories is the Judo-Christian belief described in the *Bible*, which has influenced the Christian Western world and its culture. According to the *Bible*, God created the universe in six days. On the first day, He created light and separated it from darkness. On the second day, He created an expanse and separated water from it, and called it 'sky.' On the third day, He created Earth and vegetation. On the fourth day, He created the sun, moon and stars. On the fifth day, He created all animals; and on the sixth day, He created humans in his own image and found that it was very good. On the seventh day, He took rest. According to the *Bible*, the universe is only about 6,000 years old.

Eastern Views of Creation

Hinduism

There is no concrete description of creation in Indian scriptures such as the *Vedas* and *Upanishads*. In Rig Vedic hymns, Aditi is considered the mother goddess, and mother of all gods. In another part of the hymns, there is a mention about 'hiranyagarbha'—the primeval or golden egg. Brahma is mentioned as the creator of all gods and heavens, and he is the hiranyagarbha. The *Rig Veda* takes an agnostic stance about creation, of which the following lines are evidence:

"Who verily knows and who can here declare it, whence it was born and whence comes this creation.

The gods are later than this world's production, then who knows whence it first came into being.

> *He the first origin of this creation, whether he formed it all or did not form it, whose eye controls this world in highest heaven; he verily knows it, or perhaps he knows not."*

In general, there are three schools of thought in Hindu philosophy—Advaitha (monism), Dwaitha (dualism) and atheism. Advaitha and Dwaitha believe in God as the creator. The former believes that God differentiated himself into innumerable souls, and into primeval matter and manifested himself as the world and its beings. 'Maya' or illusion clouds the beings' minds and makes them feel that they are separate from God. They go on a cycle of birth and rebirth, until they understand their true nature and get released from their material past. They then merge with God, the 'Universal Consciousness.'

In Dwaitha (dualism), the individual souls are said to be separate from God. They are also eternal. During the creative cycle, the souls unite with the elements and 'prakrithi' (nature) and manifest as limited 'purushas' or 'jeevas' (living beings). Maya clouds them. The souls continue to be separate even after the creative cycle is completed.

According to the Indian theological concept, there is a cycle of creation and destruction. The end of the cycle is called 'kalpa.' The three main divisions of time employed in the scriptures are 'yugas', 'manvantaras', and 'kalpas.'

There are four yugas, which together extend to 12,000 *divine* years (one year of mortals is equal to one day of the gods). As 360 is taken as the number of days in the year, the respective duration for each yuga is as follows:

Kritha Yuga is 1,728,000 mortal years = 4,800 divine years

Treta Yuga is 1,296,000 mortal years = 3,600 divine years

Dvapara Yuga is 864,000 mortal years = 2,400 divine years

Kali Yuga is 432,000 mortal years = 1,200 divine years.

One mahayuga, or Great Age, includes the four lesser yugas. A thousand such mahayugas are a day of Brahma, and his nights are of equal duration. Therefore a kalpa of Brahma extends over 4,320,000,000 ordinary years. Within each kalpa, fourteen manus reign; a manvantara, or period of a manu, therefore, is consequently one-fourteenth part of a kalpa, or a day of Brahma.

Kalpa means eon, and is about 4.32 billion years. It is mentioned in the *Puranas* as one day of Brahma or one thousand mahayugas. Two kalpas form a day and night of Brahma. One year of Brahma is 360 such days. Such a century form the cycle of the universe. According to Indian scriptures, about fifty such years have elapsed and there is another fifty years of Brahma before the annihilation of the universe.

This concept is very similar to the theory of Oscillating Universe—the presently expanding universe, after reaching a maximum, will start shrinking to singularity. Again, a Big Bang happens and the universe starts expanding. Accordingly, creation and destruction will go on regularly.

The four yugas form the mahayuga. Right now we are in the last yuga or Kali Yuga. After the last destruction and creation, about 453 mahayugas are over. This means that about 1.96 trillion years have passed since creation. According to scientific calculations, about 13.7 to 15 billion years have passed since the Big Bang.

Chinese

According to the Chinese philosophy, there is 'Yin' and 'Yang', the dark and the light, as the creative powers. These sustain life and living beings; they are the forces behind nature. Yang is like a dragon—hot and fiery, male, with abundant energy. Yin is like a cloud—moist and cool, female, drifting slowly. They come together for balance and harmony.

Intelligent Design

Proponents of intelligent design, such as Michael Behe, believe that the cellular mechanism is so complex that it is not possible that life originated and progressed by chance. The scientific belief is that living organisms emerged by chance from chemical reactions and interactions in the primordial soup. According to the scientific community, except for a miniscule few, intelligent design theory is a form of creationism, which is pseudo-scientific and projected as a scientific theory. The concept of intelligent design began (or got its thrust) in 1984 from the book *Mystery of Life's Origin* by Charles B. Thaxton, a chemist and creationist. This school's arguments are based on two characteristics of life—'irreducible complexity' and 'specified complexity.'

Matter Wriggles and Life Emerges

Irreducible complexity was explained by Michael Behe in his book, *Darwin's Black Box*. He defines it as, "A single system, which is composed of several well matched interacting parts that contribute to the basic function, wherein the removal of any one of the parts causes the system to effectively cease functioning." He has referred to the mousetrap as an irreducible complexity. Other examples are the flagellum of E. coli, cascade of coagulation of blood, cilia, immune mechanism and so on.

Specified complexity is the term used by Charles B. Thaxton. An alphabet is a single letter and hence not complex, but a long sentence is complex and specified. Likewise, the molecular arrangement in a DNA strand is specific and complex. A meaningful sentence cannot come into being by chance. Likewise, a meaningful arrangement of molecules cannot happen by chance. The proponents of intelligent design put forward this hypothesis.

Richard Dawkins observed that the designer would also have to be equally complex. Naturally, there should be a designer for the designer as well, and that means that the problem of origin of life or origin of meaningful complexity is only postponed and not solved. Dawkins sees this hypothesis as a "thought terminating cliché."

Another argument put forward by intelligent designers is that the laws of the universe, like gravity, nuclear force and so on are meant to support life. If there were a slight change, life would not have existed. Scientists, though, agree that the problem is not fully resolved, and believe that the origin of life and evolution can be better explained by natural selection and not by intelligent design

To say that other forms of life are not possible implies a lack of imagination. Jerry Coyne stated, "Either life resulted not from intelligent design, but from evolution, or intelligent design is a cosmic prankster who designed everything to look as though it had evolution." Behe's argument gives only negative arguments against evolution, and does not provide any positive argument for intelligent design. The function of vestigial organs and the reasons for not producing the best design at the first instance and so on has been explained as "unknowable motives" of the designer. The scientific community unambiguously rejected the concept of intelligent design, as it did not qualify the 'requirements of a scientific theory.'

Scientific Concept of Origin of Life

Is life is a property of matter?

For life (as we know of it today) to originate in a planet or in a moon of a planet, there are certain requirements. The star should be in a habitable zone of the galaxy, and the planet should be in the habitable zone of the star. The place should not be too hot or too cold. The planet that can sustain life is also known as a 'Goldilocks planet' (a term taken from the children's story, 'Goldilocks and the Three Bears'). The planet should not be too close or too far away from the galactic center or star. The temperature should be just right for water to exist in liquid state. The size and lifespan of a star determines whether intelligent life can originate in a planet. Elements like carbon, nitrogen, oxygen, phosphorous, and so on, in the right proportion, which are the building blocks of life, are also essential for the origin of life. The size of the planet, its orbit and rotation, and its proximity and distance from other planets, and so on, are all factors that influence the development of life. Because of this multitude of stringent requirements, many consider the presence life to be quite rare in the universe. However, a rough estimate is that there will be at least 500 million planets that are probable candidates for the formation of life in the Milky Way, our own galaxy. The nearest such planet is about twelve light years away.

Before we consider how life originated, we have to understand what life is. Science was not able to decipher life in an absolute sense. A living organism is a complex association of non-living material, which in unison expresses the mysterious property called 'life.'

Though incomprehensible, certain characteristics and properties of life are known to science. An important one is its dynamic and self-sustaining nature. Living organisms grow and reproduce, undergo metabolism and respond to stimuli. As a response to external stimulus, an organism adapts to its surroundings—the primary cause to become a better organism by 'natural selection.' Korn and Korn (1971) listed a few characteristics that differentiated living organisms from inanimate matter. They are: capacity for synthesis, capacity to self-regulate and capacity to adapt genetically (evolutionary change). One of the important (probably the most important) functions of any organism is reproduction. This can be paraphrased as 'survival by transfer of genes', as Richard Dawkins has put forth in his book, *The Selfish Gene*.

There are certain prerequisites for any life form to reproduce. The organism has to grow—and, for growth, it has to assimilate and convert inanimate material to living parts of itself and discard what is not required. This mechanism is called 'metabolism.' A living organism provides all the information, catalysts, enzymes and proteins for reproduction and manufactures these essential ingredients from its surroundings. Viruses do not have the catalysts necessary for reproduction. They burrow into a living host cell as a parasite and reproduce using the host cell's facilities. There is no definitive proof for the spontaneous origin of life from non-living material. There are only possible and plausible hypotheses and postulates. Some scientists believe in the theory of creation by a supreme being. They might have been driven by the lack of evidence for abiogenesis. A majority of the scientists are agnostics—attributing the origin of life to a creator is against the spirit of scientific quest. So, the search continues.

Spontaneous Generation

Early philosophers like Aristotle and his contemporaries believed that organisms spontaneously originated from non-living matter. They believed that aphids originated from dewdrops, flies from putrid material, bees from flowers and crocodiles from decaying logs in water.

In 1676, Anton Van Leeuwenhoek, popularly known as the 'father of microbiology', invented the microscope. His contribution to microbiology was noteworthy, and it helped the progress of biological sciences. Francesco Redi's (1626–1698) and Louis Pasteur's (1822–1895) experiments undoubtedly disproved the concept of spontaneous regeneration.

In 1871, Charles Darwin famously wrote a letter to Joseph Dalton Hooker, suggesting that life might have started in a warm little pond with all sorts of chemicals, such as ammonia and phosphoric salts. Light, heat and electric discharges might have acted on the compounds to form a protein, which later underwent more complex changes.

Panspermia

This hypothesis argues that life was seeded on Earth by aliens, or that it might have taken root here from meteorites or comets. Panspermia does not answer the question of how life originated.

Richter (1865) developed this theory, and he had the support of many scientists. During the early days of its existence, Earth was pelted by meteorites in ample numbers. As they passed through the Earth's atmosphere, their outer surface became red-hot but the inner core might have remained cold—and this might have harbored the seeds of life. Panspermia may explain the origin of life on earth, but doesn't explain the origin of life in the universe.

Conditions similar to Earth's atmosphere might have been present in many areas of the solar system. This is evinced by the presence of more than ninety amino acids in the Murchison meteorite, which fell near Murchison (Victoria, Australia). Of the ninety, nineteen are found in earthlings. Early Earth was pelted by meteorites that contained many organic compounds and amino acids, and primitive life forms might have been created through those meteorites. Panspermia hypothesis depends on these findings. Francis Crick, co-discoverer of DNA, a Nobel laureate, supports this theory. Robert Shapiro (1935–2011) of the New York University opined, "Life is a normal consequence of the laws of nature and potentially quite common in the universe."

Primordial Soup

Alexander Oparin, a soviet biochemist, wrote the book *The Origin of Life on Earth* in 1924. He proposed that complex combinations, manifestations and properties characteristic of life must have arisen in the process of evolution of matter. This means that 'life is a property of matter', which is exhibited under proper conditions.

During the early days of the Earth's formation, it was a reducing atmosphere due to the lack of oxygen. There was methane, ammonia and water vapor—the raw materials required for the origin of life. There is a tendency for most chemicals to combine and form stable compounds. Oparin has applied Charles Darwin's concepts of struggle for existence and natural selection into the material world as well, and was considered as the 'Charles Darwin of the 20th century.'

It is apt to quote John Burden Sanderson Haldane at this point: "It seems to me immensely unlikely that [the] mind is a mere byproduct of matter, for if my mental process is determined wholly by the motions of atoms in my brain, I have no reason to suppose that my beliefs are true. They may be sound chemically. And hence I have no reason for supposing my brain to be composed of atoms." (J.B.S. Haldane was a British-born geneticist and evolutionary biologist and staunch Marxist, who moved to India and became an Indian citizen. He was one of the founders of population genetics.)

In 1929, Haldane proposed that ultraviolet rays, acting on a primitive atmosphere containing methane, ammonia and water vapor, produced oceans with the consistency of a hot, dilute soup containing the building blocks of life. The process of evolution of living matter from these building blocks is called 'biopoesis.' Due to the action of sunlight on the primordial soup, organic molecules were created in an oxygen-free atmosphere. These complex molecules produced coacervate droplets, which were tiny and spherical organic molecules of assorted nature. Hydrophobic forces held them together. In water, organic molecules aggregated and formed droplets surrounded by a tight skin of water molecules. They absorbed simple organic molecules from the surroundings. Oparin considered this as a simple form of metabolism. Both Oparin and Haldane were of the opinion that evolution is a purely chemical process, and that organic molecules developed from inorganic molecules.

Bernal, known for his pioneering work in crystallography and molecular biology, opined that coacervates are "...the nearest we can come to cells without introducing any biological or at any rate, any living biological substance."

Miller-Urey Experiment

Most amino acids—the building blocks of life—can be formed through chemical reactions. An experiment by Stanley Miller and Harold Urey, conducted at the University of Chicago in 1952, was a turning point in the search for the origin of life. (Miller and Urey published their results in 1953.) They half-filled a flask with ammonia, water and hydrogen and boiled it. The vapor went into another flask, where there were electrodes and sparks. This was to create a pre-biotic atmosphere. After a week of continuous operation, they were able to get more than twenty amino acids and sugar. Many others repeated the experiment and got the same results. Even nucleotides like adenine were produced in these experiments.

Sidney W. Fox, through his (near similar) experiments, was able to produce not only protenoids but also small cell-like structures, which he called microspheres (proto-bionts). Though they were not proper cells, they could form double membranes, which could perform osmosis. However, they had no DNA.

Replication of the molecules gives rise to a riddle: What is replicated first—the proteins or the nucleic acids? The 'RNA world' hypothesis proposed that RNA molecules were formed by the polymerization of nucleotides and resulted in self-replicating ribosomes. Ribonucleic acids carried genetic material and could catalyze chemical reactions. They produced proteins, which have better catalytic activity than ribosomes and became the dominant biopolymer.

Joan Oro was able to synthesize cytosine, uracil and adenine—the bases for DNA—from a urea solution. 'The Soup Theory' is based on the assumption by Charles Darwin that chemical evolution occurred in the environment before life originated. Self-organization and replication are two important characteristics exhibited by living organisms. The RNA world hypothesis describes that there was self-replicating and catalytic RNA during the early years on Earth.

In 1993, Stuart Kauffman proposed that life initially arose as autocatalytic chemical networks.

There are several and varied concepts about the origin of life. Several forms of life might have appeared during the same period and become extinct later, except for the one that was most appropriate.

The origin of proto-cells could be considered as the preliminary stage in the formation of living organism. When proto-proteins came in contact with water, they underwent self-assembly and formed proto-cells, which had a spherical structure with a double-layered membrane with semi-permeable properties. These cells were not the result of reproduction, but were formed through the combination of available organic molecules in the environment. The beginning of life, though in the primitive form, was through the self-assembly of polypeptides into proto-cells, which had the capacity to replicate. Of these proto-cells, many did not survive—which means that there was natural selection. Those that survived could incorporate adenosine triphosphate (ATP), pyrophosphate or polyphosphate utilizing mechanisms, which are the basis for metabolism. The cells developed the capacity to synthesize proteins, including nucleic acids and ribosomes. Coded macromolecules such as DNA—the genetic material that could orchestrate the entire range of functions of the cells—appeared, and the present form of cells emerged. Thus, present organisms emerged through natural selection.

Also, known as the essence of life, DNA produces proteins—an inevitable part of life. Duplication of DNA is essential for replication of cells; DNA polymerase is the enzyme (a protein) that acts as a catalyst for DNA production. So, DNA could not have originated first. In other words, a paradox exists regarding the origin of life—DNA is essential for production of proteins, and proteins are essential for production of DNA.

In the early 1980s, Thomas Cech and Sydney Altman discovered certain forms of RNA called ribozymes, which could act as an enzyme. So, many believe that RNA came first (before DNA). It is estimated that these primitive forms took about one million years to progress into a self-replicating life form.

To maintain stability, a protective mechanism had to be put in place, and a membrane formed—which is the cell wall. A repair mechanism had to be in place should any injury happen. The semi-permeable nature of the cell membrane allowed transport of small repair units available in the environment, and the excess could be stored inside for future use.

Considering the origin of matter and its progress from basic particles to complex molecules, to atoms and through molecules to complex compounds, an evolutionary trend for stability and self-organization can be envisaged. A free hydrogen atom cannot exist by itself—it shares its electron with another atom and forms a molecule for stability, which is notated as H_2.

Endosymbiosis

The Endosynthetic theory of Lynn Margulis proposed that multiple forms of bacteria entered into a symbiotic relationship to form the eukaryotic cell—the 'Last Universal Common Ancestor' (LUCA) of all modern organisms. Prokaryote by definition is a microscopic single-celled organism, which has neither a distinct nucleus with membrane nor specialized organelles. This includes the cyanobacteria. Eukaryote by definition is an organism consisting of a cell or cells in which the genetic material is DNA in the form of chromosome contained within a distinct nucleus. Eukaryotes include all living organisms other than the eubacteria and archaea.

Eukaryotes are considered to be the decedents of prokaryote cells. The conventional concept was that the prokaryote cell had genetically mutated to a complex form to become eukaryotes. Lynn Margulis brought out the theory of Endosymbiosis in 1960. It was not accepted initially; in 1981, she published a book with strong arguments, and the scientific community accepted her theory of endosymbiosis.

Mitochondria, chloroplasts and other organelles such as lysosome and flagella were prokaryotes. Mitochondrion is an organelle found in most cells in which the biochemical processes of respiration and energy production occur. It has a double membrane, the inner part being folded inwards to form layers (cristae). Green plants' cells have chloroplasts—a plastid that contains chlorophyll, in which photosynthesis takes place. Lysosome is an organelle in the cytoplasm of eukaryotic cells, containing degradative enzymes enclosed in a membrane. They can lyse most of the biomolecules. It is a waste disposal system of the cell. Flagellum is a slender thread-like structure, a microscopic whip-like appendage. These enable many protozoa, bacteria, spermatozoa, and so on, to swim. These prokaryotes entered large anaerobic cells and received nutrients. The anaerobic host cells, in turn, received benefits from these prokaryotes, such as photosynthetic cells and mitochondria. These organelles helped the anaerobic cells to survive when oxygen increased.

Mitochondria and chloroplast have their own DNA without a nucleus. Through endobiotic gene transfer, some of the genes were given off to the host cell and depended on the host cell for synthesis of components. Mitochondria are responsible for the energy production for the cell. However, some waste is produced in this process and these free radicals are supposed to be a reason for the ageing of cells.

(In humans, mitochondria are inherited maternally. Research for the 'mitochondrial Eve' ended around two hundred thousand years ago in Africa. This is still a subject of debate.)

Life started about three-and-a-half to four billion years ago. A.I. Oparin and J.B.S. Haldane were supposed to be the pioneers who proposed the concept of origin of life by natural methods from non-living chemicals of the primordial soup. Pioneering work by Miller and others were able to produce building blocks from non-living substances. The discovery of DNA and RNA, their catalytic activity, self-assembling lipids forming membranes and so on suggest the emergence of life by natural selection.

2
Life Tickles/Perception

Evolution of nervous system

Plants are auto-tropic and there is no need for them to move around. However, animals have to move around. Hence, they face different environmental situations and have to adapt to maintain homeostasis—a steady state of the body. Hence, two coordinating mechanisms present in animals are the nervous system and the endocrine system.

The evolution of the nervous system in animals is an exhilarating story. From the action potentials in single-celled organism to nerve nets in lower forms of animals, especially coelenterates, nerve cords in bilateral animals, ventral nerve cords in invertebrates, dorsal nerve cords in chordates and later cephalization (formation of the brain) is in short the story of evolution of the nervous system.

An amoeba does not have a nervous system—its movement is by molecular mechanism. It is a crawling type of movement putting forth pseudopodium (a projection on the surface of the cell). This movement is by change in the action potential (change in electrical potential). Amoeboid movements are seen in slime molds, protozoans and so on. (Human leukocytes also show this type of movement. Malignancies arising from connective tissue [sarcomas] are proficient in this type of movement, which is one reason behind their high metastasis.)

Chemotaxis is the movement of an organism due to chemical stimulus. It could be positive (towards the chemical) or negative (away from the chemical). This is an important method of movement of single-celled organisms. Movement of sperm cells towards the egg is an example of chemotaxis (movement due to chemical stimulus).

Action Potential

Action potential is the beginning of transmission of energy. This happens in excitable cells like neurons, muscle cells and endocrine cells. Exchange of ions (mainly Na^+ and k^+ ions.) in and out of axons happens by polarization and depolarization. The action potential travels from the neuron down the axon to the terminal, where it signals another neuron or muscle. Action potential is present in single-celled eukaryotes.

Evolution of the nervous system is linked to the development of voltage gated sodium channels, which allows long distance transfer of stimulus. Calcium channels allow intercellular signaling. Sodium channels are advancements over calcium channels. This system is used by porifera-sponges.

Nerve Net

Interconnected neurons, lacking a brain, have action potential transfer through nerve nets. Organisms that have radial symmetry (commonly known as coelenterates) have nerve nets. (Radial symmetry—they have a top and a bottom; they are grouped into taxon radiata-like cnidarians [such as hydra], echinodermata [such as sea stars, sea anemone, jellyfish and so on].)

The nerve net, the simplest form of nervous system, allows the organism to respond to external stimuli but cannot detect the source of stimulus (they produce the same motor output irrespective of the point of contact). The neurons are not grouped together as in a nervous system. Cnidarians have a nerve net that extends throughout the body. The echinoderms have nerve nets in each arm and a central nerve ring that connects the nerve nets to the center.

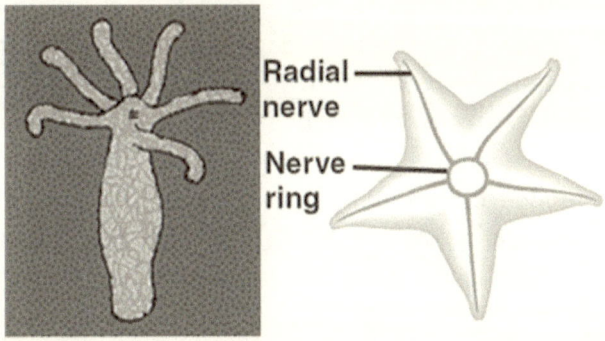

Fig 2   *Nerve net in hydra and the nerve ring of starfish*

Sponges are animals belonging to the phylum porifera (meaning 'pore bearer'). They are a conglomeration of unspecialized cells, which can transform to other types. The organism is full of pores and channels, which allow water to flow through them. They don't have circulatory, digestive or nervous systems. They rely on the constant flow of water for their nutrients, oxygen and disposal of waste. Most species of sponges can move, which is a coordinated activity. Though the mechanism of this coordination is unknown, it is presumed that some neurotransmitters are responsible.

Cnidarians have two layers of tissue, ectoderm and endoderm, separated by an extracellular matrix called mesoglea. Interstitial stem cells from the endoderm gives rise to neroblasts, which migrate to the ectoderm and give rise to the nervous system. This axis provides light to the divergence of coelenterates and bilaterians.

Nerve Ring

Echinoderms are (usually) five-point radial symmetrical marine animals. The well-known ones in this group are the starfish, sea urchin, sand dollar and sea cucumber. They have a remarkable capacity for regeneration and asexual reproduction (some species reproduce sexually as well). Sometimes the entire individual may be regenerate from a single limb. Though larvae of echinoderms are bilaterals during metamorphosis, they acquire radial symmetry. Probably the ancestors of all echinoderms were simple bilaterally symmetrical organisms with a mouth, gut and anus.

The organism's nervous system is reasonably simple. The nerves radiate from a central nerve ring around the mouth to the arms along the body wall. These branches coordinate the movements. Epithelium of the starfish has sensory cells, touch-sensitive tentacle-like tube feet at the tip of the arms, and simple eyespots. Each eyespot is made of eighty to 200 simple ocelli, which are composed of light-sensitive pigmented epithelial ells.

The thick transparent cuticle protects the ocelli and helps the starfish to focus. Starfish also have photoreceptor cells in other parts of the body, to which it can react.

The nervous system is complex, and starfish have sensory systems at the epidermis and a motor system in the lining of the coelomic cavity. These two are connected by neurons passing through the dermis.

Sexual reproduction starts when they become adults (when they are about three years old). The sperms and eggs are spewed into the water, and fertilization takes place.

Nerve Cords

Nerve cords are an advanced development over the nerve net. This is seen in bilaterians—animals that have a symmetrical body, with left and right sides being almost mirror images. (Bilaterians are supposed to have descended from a worm-like ancestor, around 550 to 600 million years ago.) Invertebrate bilaterals have a ventral nerve cord, while the vertebrates have a dorsal nerve cord. Worms are the simplest form of bilaterians. The earthworm has a dual nerve cord that is connected like a ladder by nerves. These transverse nerves help coordination of both sides. At the head-end are two ganglia that function like a brain. Photoreceptors are present on the eyespots.

Arthropods, like insects, also have a nerve cord, with each segment having a ganglion. They have a brain and a ventral nerve cord. At the head-end, three pairs of ganglia fuse to form the brain, which is otherwise known as supra-oesophageal ganglion.

The brain has three parts—protocerebrum, deutocerebrum and tritocerebrum. The number of ganglia varies for different arthropods.

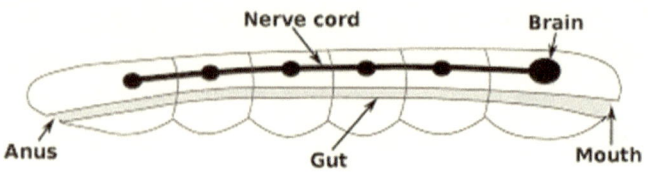

Fig 3   *Arthropod nervous system*

The sensory systems of most arthropods are well developed, with compound eyes and antennae for olfaction. Sensations are processed by the brain. The sub-esophageal ganglion, which comprises three pairs of ganglion, controls the movements of the mouth, salivary glands and certain muscles.

A study by Tracey J. et al in 2003 pointed out that insects may have nociceptors, and feel pain. Many insects have specialized organs for perception. Bees can perceive ultraviolet and polarized light. Moths can detect the pheromones of females from kilometers away. Insects that can produce sound can hear as well.

Arthropods have compound eyes. Most arthropods have lateral compound eyes and a median ocelli (little eyes)—the simple form of eyes. Ocelli can only detect the direction of light. A compound eye consists of several thousand columns, hexagonal in cross-section, and is an independent sensor with its own lens and cornea. Arthropods are often short sighted (about twenty centimeters). Many of them have color vision. Bees can detect both green and ultraviolet light.

Vertebrate Brain Evolution

Most vertebrates have the same divisions in their brains. The relative size of the brain in comparison to the body, as well as the size of each part of the brain, varies from species to species. Though brain size increases with body size, this is not proportional but allometric (differential growth of body parts). The variation in brain size is in an orderly fashion.

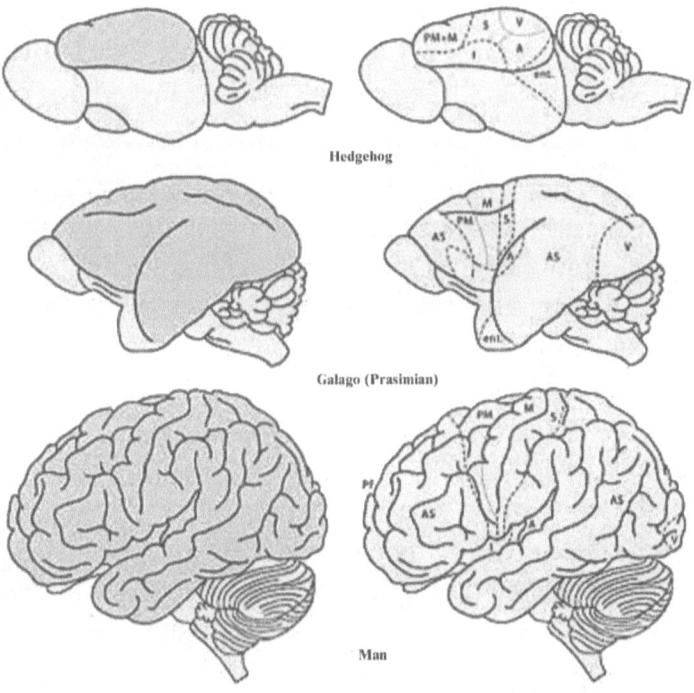

Fig 4 *Vertebrate brain evolution*

Agnathans are a group of jawless vertebrates. They include lampreys, hagfishes and many fossil fish forms.

They lack cerebellum, but they possess a thin layer of tissue adjacent to the medulla, which is interpreted as a primitive cerebellum (Larcell, 1967). These organisms have the smallest brain among vertebrates. Hagfishes have brains that are two to three times larger than that of lampreys of the same body size.

Most of the bony fishes have a larger brain size than agnathans of same body size. Among fishes, teleosts such as ray-finned fishes, lungfishes and coelacanths have a larger brain than non-neopterygians such as paddlefish and bichris. However, they have smaller brains than amphibians. Among amphibians, frogs have larger brains than salamanders. The brains of reptiles are about three times larger than that of amphibians.

Birds have brains that are about six times larger than reptiles relative to their body size. Among birds, the largest brains are found in perching birds such as woodpeckers and parrots. Pigeons and chickens have smaller brains

Mammals' brain sizes are ten times that of reptiles with the same body size. Of the mammals, primates and cetaceans (whale, dolphin) have the largest brain relative to their body size. Non-placental, marsupials, insectivores, rodents and so on possess the smallest brain among mammals. It is not easy to understand the selective pressure of variations in the brain size.

Features of the Nervous System

For the brain to get information about the site of stimulation, each fiber should be different. There are two types of fibers that transmit pain—myelinated and demyelinated. The speed at which the former transfers the information is about five to thirty meters per second, while the demyelinated transmit at a speed of one meter per second. Often, sharp pain is transmitted fast and as a reaction by reflex action. The slower one is a continuous ache, and acts as a warning.

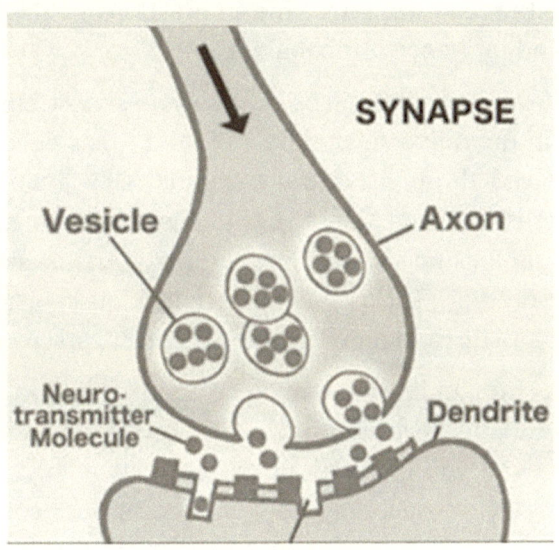

Fig 5   *Neural synapse*

Nerve cells have dendrites—which receive information, and axons—which send out instructions or information. Dendrites are short and highly branched (dendrite means 'tree' in Greek). Axons are long and branched at the end. The end of the axon is expanded like a knob and is in contact with a dendrite of another cell. This contact point is the synapse ('connection' in Greek). A chemical transmitter at the synapse in the neuro-muscular junction is acetylcholine. Synapses at the central nervous system could be excitatory or inhibitory. Excitatory transmitters are glutamate and inhibitory transmitters are gamma amino butyric acid (GABA). Interactions between the nerve cells influence our thinking, feelings (qualia), learning and memory. Drugs, such as sedatives, hypnotic analgesics, tranquillizers, anxiolytics, mood elevators, hallucinogens and antipsychotics, affect the interactions of the nerve cells and influence our behavior.

The nervous system is probably the most complex thing in the universe we know of, and it has developed or evolved through natural selection. Development of consciousness is a subject of enthusiastic debate among philosophers and scientists from the pre-Aristotelian era to the present. The popular concept, 'Consciousness is characteristic of higher animals', is controversial.

The human brain has about a hundred billion (10^{11}) neurons, and each neuron has more than a thousand synapses. If you counted all the neurons in your brain at the rate of one a second and never lost count, it would take 645 years to count them all! The number of potential connections that could be made between neurons is greater than the number of known atoms in the universe! You could fit 30,000 brain neurons on the head of a pin! However, other neurons can be several feet long. For instance, the length of a giraffe's longest neuron (from its toe to its neck) is fifteen feet!

A widely accepted concept is that humans have five senses. In fact, we have more—such as awareness of our body parts and ourselves, sense of balance and so on. Many animals have certain senses that we don't possess, and some of them are superior to ours. For example, bats have the sense of echo locomotion. Migratory birds have sensitivity to the Earth's magnetic field. Some fishes can sense changes in the electrical field.

Development of the Brain

The nervous system starts as a midline groove in the embryo at the dorsum, which deepens into a tube that later develops into the spinal cord and brain.

By the fourth week of gestation, the neural tube enlarges and curves forward and forms three bulges. The front end is the prosencephalon, which later becomes the forebrain. At the basal plate of the prosencephalon is the optical vesicle, which develops into the optic nerve, retina and iris. The middle part is the mesencephalon— which develops into the midbrain, and the hind part is rhombencephalon— which develops into the hindbrain.

Forebrain

During the evolutionary process from fish to amphibian, reptile and mammal, there is progressive enlargement of the endbrain. The endbrain has two cerebral hemispheres, which gradually enlarged as evolution progressed from the lowly mammal to primates and humans. The surface of the cerebral hemispheres has a layer of nerve cells, which is about two to five millimeters thick. This layer is the cerebral cortex, and is known as the 'grey matter.' Though the surface of the cerebral cortex is only a quarter of a square meter, the grey matter has got around a hundred billion neurons.

The bundle of fibers under the cortex is the axons, and this is the 'white matter.' The bundle of axons crosses the hemisphere to the opposite one—and is known as corpus callosum. Information is transferred from one side to the other through the corpus callosum.

Deep in the white matter are the basal ganglia, hippocampus and amygdala, which have neural collections. The basal ganglion is responsible for the control of movements. The hippocampus resembles a sea horse in shape (hence this name), and the amygdala has an almond shape. They are concerned with memory and emotions.

All these senses except smell pass through the thalamus to the cortex. The hypothalamus, which is below the thalamus, controls the hormonal system through the pituitary gland and acts through the autonomic nervous system—sympathetic and parasympathetic. The former controls blood pressure and the latter controls most of the other involuntary functions such as appetite, thirst, salt and water balance, body temperature, movements of the gut and so on. The sympathetic nervous system prepares the body for vigorous action—fright, flight and fight. Sleep and wakefulness is controlled by the hypothalamus in collaboration with other parts of the brain.

The pineal gland is a prominent part at the center of the brain between the two brain hemispheres, where the two halves of the thalamus meets. The function of the pineal gland is unknown, and is considered as a vestige of the third eye seen in the fossils of some primitive fishes that are extinct. Tuatara, a living reptile found in New Zealand, possesses this third eye. In certain mammals, the secretion (melatonin) of the pineal gland is involved in maintaining circadian rhythms (a rhythm which follows a twenty-four-hour cycle).

Midbrain

The midbrain in fish, amphibians and reptiles (vertebrates other than mammals) is one dome on each side—known as the optic lobe, since it is concerned with vision. In mammals, the roof of the midbrain has four lobes, corresponding optic lobes of other vertebrates, concerned with the movements of the eye and with auditory information before it is transmitted to the cerebral hemispheres. In bats, it is relatively large and is involved with echolocation. The midbrain collaborates with the forebrain to control movements, sleep and arousal.

Hindbrain

This part connects the midbrain with the spinal cord. The front part is larger and folded, and this is the cerebellum—which has to do with posture and balance. In all vertebrates, the cerebellum controls circulation and respiration (both of which are involuntary functions). In lower vertebrates, this is concerned with the senses of vibration and taste. In mammals, these functions are taken over by the cerebral cortex.

Cerebral Cortex

The surface of the human cerebral cortex is convoluted. The fissures on the surface of the brain—Fissure of Rolando and Fissure of Sylvius divide the hemisphere into frontal, parietal and temporal lobes. The occipital lobe is at the back of the brain. The surface of each lobe has numerous gyri (a ridge on the cerebral surface of the brain) separated by smaller fissures, which makes the surface of the brain look convoluted.

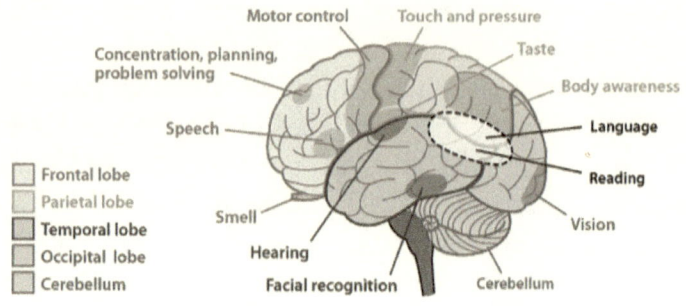

Fig 6 *Functional areas of the brain*

Lots of experiments were conducted to find the function of each part of the brain. From the time of Hippocrates, it was known that the brain controls the opposite side of the body. He warned against making incisions on the brain as it causes convulsions on the opposite side of the body. Hughling Jackson, Gustav Fritsch, Edward Hitzig, David Ferrier, and Paul Broca are important people who have contributed to the knowledge of the function of the brain.

3

Cognition and Confusion

Evolution of the Brain

Brain size increased from fish to mammals, from the oldest taxonomy groups to the recent ones. The brain has different functions, which are controlled by different systems. The human cortex is twice as thick of as that of mice and has a thousand-times larger surface area. The human brain surface is ten times more than that of macaque monkeys and thicker by fifteen percent. We share about ninety-eight percent of our genes with them. The increase in surface area is attributed to the increased folds, which can hold more number of neurons and more connections. Neurons cover the surface area of both hemispheres.

The importance of the cerebral cortex has improved multifold since the appearance of mammals about 200 million years ago. The brain started enlarging in mammals, mainly in the cerebral cortex. Between the herbivores and the predatory carnivores, the brain enlargement was more in the predatory mammals and, that too, in the neo-cortex. Predation was more difficult than grazing; meat is more nutritious and provides more calories and energy.

Brain enlargement occurred in certain areas and by addition, rather than by total reorganization or remodeling. The early parts of brain catered to fundamental needs. The complexity of the social life of primates could be the driving force behind the rapid expansion of the neo-cortex in primates. The growth of the neo-cortex favored the enhancement of social skills such as language. Increase of the cortical surface was accomplished by the increase of folds on the surface of the brain. Complex social interactions, hunting, storing, sharing and awareness are attributed to having a big brain.

Nature, (possibly) on its path towards perfection, experiments with various life forms, through bizarre evolutionary processes—there may be failures (which will become extinct in due course), and successes (which will continue to survive and even diverge into better forms). Though the brain size in humans is large, man cannot be considered a special creation, as other existing life forms continue to flourish in harmony with nature. A major threat to them, in fact, is the exalted human.

Brain size alone cannot be considered as the reason for intelligence. Whales and elephants have brains that are about four to five times larger than that of humans, and their body size is much larger. Proportion between the body size and the size of the brain may be taken in to consideration. However, there are certain exceptions—hummingbirds and certain squirrels in Central America have larger brains than humans in proportion to their body size.

The size percentage of the cerebellum remains almost constant in all mammals in relation to the total brain size. Fishes and amphibians do not have a neo-cortex. Shrews' neo-cortex account for twenty percent while that of humans is about eighty percent of the brain. During the evolution from primates to hominids, the expansion of the neo-cortex, especially the prefrontal cortex, was rapid.

Three parts of the brain evolved successively:

1) Archicortex, composed of dentate gyrus and hippocampus in mammals, and is associated with the olfactory system

2) Paleocortex, composed of pyriform cortex and the para hippocampal gyrus—they are also associated with the olfactory system

3) Neo-cortex (isocortex) is a recent appearance—it is stratified into six distinct layers of neurons and is associated with higher functions like intelligence

Brain Evolution Model

Paul MacLean has proposed the triune brain theory to explain the evolution of the brain in his book, *The Triune Brain in Evolution*.

1) Reptilian brain (archipallium or the primitive brain)

2) Limbic brain (paleopallium or intermediate, known as the 'old mammalian' brain)

3) Neo-cortex (neopallium, also known as the superior or rational, 'new mammalian' brain)

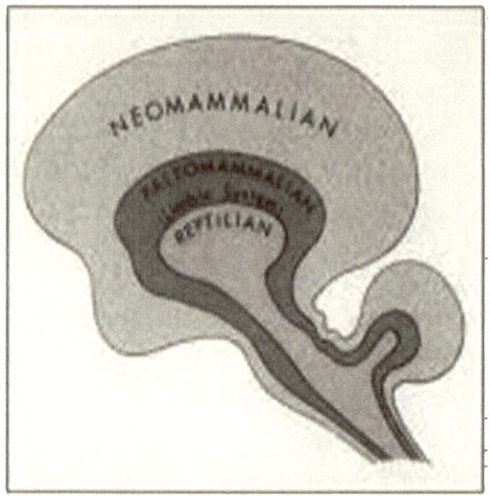

Fig 7 *Evolutionary model of brain; triune brain*

Each part of this has been added subsequently during evolution. According to Paul MacLean, each part of the brain has "their peculiar types of intelligence, subjectivity, sense of time and space, memory, mobility and other less specific functions." MacLean conducted extensive studies and proposed his triune brain theory to explain the evolution of the brain and rational behavior of humans. His human study was on patients suffering from brain disorders such as epilepsy, and he later extended his study to cats and other animals. His theory is widely accepted in neuroscience.

Scientists such as Joseph E. LeDoux agree that the limbic system has a strong influence on emotions. LeDoux conducted research on rats to find the origin and architecture of emotions. The emotion of fear helps the animal to escape from predators, and it is more instinctive. Hence, it is difficult to control emotions by rational and conscious thought. However, LeDoux opined that it is not right to demarcate the brain by the triune system, and felt that the brain has to be thought as an 'integrated whole.'

It may not be accurate to term brain parts as 'reptilian' and 'old mammalian', as scientific studies have shown that the common ancestors of reptiles and mammals had well-developed limbic system. Birds (which evolved from dinosaurs) have a well-developed limbic system that has almost the same function, though anatomically different, which means that it is an instance of 'convergent evolution.'

Reptilian Brain

The reptilian brain controls vital functions such as the heart rate, breathing, body temperature and balance. Parts that control these functions are the basal ganglia, brain stem and cerebellum. These are structures derived from the base of the forebrain. This is responsible for specific instinctive behavior of the species, such as aggression, dominance, territorial and ritual displays for protecting the territory, and attracting sexual partners. The reptilian brain is seen in fish, amphibians, reptiles and mammals. This might have developed about 500 million years ago in the common ancestor of vertebrates.

The reptilian brain is the most ancient part of brain, comprised of the upper part of the spinal cord, the basal ganglia, diencephalon and parts of the midbrain. This is an inevitable part of the mammals' brains, as it controls vital functions of the body and is considered to be the evolutionary link between reptiles, dinosaurs and mammals. Another, and probably the primary, social function of the reptilian part of the brain is protection of the territory.

The reptilian brain is subdued during the waking hours; while sleeping, it function and dreams. (As reptiles have an active reptilian brain, they may be in a dream state during waking hours, as Carl Sagan has opined in his book *Dragons of Eden*). Infants spent most of their time in rapid eye movement (REM) sleep—a dream state. During early childhood, the reptilian and mammalian brains control them, which is important for basic needs like food and shelter. (Example: The baby that cries gets the milk.)

Sigmund Freud (1856-1939), 'father of psychoanalysis', who wrote the book *Interpretation of Dreams*, considered dreams as a method of wish fulfillment. Its entire range of actions are motivated by the unconscious. The action of the super ego censors many of our urges and impulses. These emotions are expressed as dreams in a symbolic manner. The unconscious rules during the sleep. If there is a need for the body, such as thirst, while asleep, the conscious mind tries to wake up the person so that he can drink water. The unconscious mind doesn't want to lose its supremacy. The person dreams that he is drinking water. This is a lie told by the unconscious mind to the conscious mind to satisfy the latter. For some time, this dream can dissuade the conscious mind. However, when the stimulus becomes strong, the person wakes up.

Limbic System

The limbic brain, which evolved in mammals, records memories and experiences such as emotions, value judgments and behavior, and is responsible for motivation, emotion, feeling, reproductive behavior and parental behavior. The hippocampus, septum, cingulate cortex, amygdala and hypothalamus belong to the limbic brain. Paul McLean called it the 'paleomammalian complex', since he has considered that it had been added to the brain during early mammalian evolution. The 'flight or fight' response, which is important for survival, is controlled by the limbic system, and it is older than neo-cortex.

Smaller than a dice, the hypothalamus is probably the most powerful part of the brain as it is concerned with emotions and regulates autonomous body process. The hippocampus is concerned with smell and short-term memory. It processes information to assess whether it is worth remembering or not, and sends it to other areas. Actually no site in the brain has a monopoly on memory.

The amygdala, an almond-shaped structure below the hypothalamus, is the aggression center, as its stimulation can cause negative emotions such as anger, aggression and hostility.

The limbic system is a strategic part with cross-connections from both higher and lower centers of the brain. It controls motivation and emotions. The cortex moderates emotions and motivations that originate in the limbic system. (It can be said that limbic system proposes and the cortex disposes.) The limbic system is where the subcortical structures meet the cerebral cortex, and controls endocrine and autonomous nervous systems. The brain's 'pleasure center', the nucleus accumbens, is strongly connected to the limbic system and influences sexual feelings.

The limbic system is considered a 'mini brain on the borders of reason.' In humans, the large cortex subdues this part. In lower mammals, the limbic part is relatively larger. Dogs have large olfactory connections in the limbic system, which accounts for their superior olfactory ability.

Neo-cortex (Neo Mammalian Complex)

'Neo-cortex' in Latin means 'new bark.' It is also known as neo-pallium. It developed in mammals, and is the latest evolved part of the brain. Ninety percent of the cerebral cortex is the neo-cortex. The ratio between neo-cortical grey matter and medulla is thirty-to-one in chimpanzees, but sixty-to-one in humans.

It is hypothesized that the neo-cortex evolved to suit the pressures of group living, cooperation and competition among the early hominids. Increase in the size of the neo-cortex has imparted voluntary inhibitory control of behavior, begetting better social harmony. The six-layered cortex is a unique feature of mammals. While birds also have cognitive process, they do not have six layers of neo-cortex.

The neo-cortex gained importance in primates and reached its acme in humans. Large cerebral hemispheres are the seat of the neo-cortex, involved in higher functions such as sensory perception, generation of motor commands, spatial reasoning, controls abstract thought, imagination, language, consciousness and learning abilities. These were the basic factors that helped us to develop cultures. The neo-cortex has deep grooves and ridges, which increase the surface area. Neurons are positioned on the surface. In lower mammals, like rodents, the surface is smooth.

The brain is organized as forebrain, midbrain and hindbrain. All the different parts of the brain are interconnected and operate in harmony through neural pathways. The neural pathways between limbic system and cortex are very strong.

White matter, composed of the myelinated axons, is meant to communicate to other parts of the brain. A large volume of white matter is situated in the prefrontal cortex. These connections are vital for memory. Brain imaging has shown that the prefrontal cortex is active during memorization.

The neo-cortex developed from the telencephalon, which is the front part of the forebrain. The neo-cortex has frontal, parietal, occipital and temporal lobes. The temporal lobe is concerned with audition; the occipital lobe is the primary visual cortex. In humans, the frontal lobe is concerned with cognition, language (Broca's area), social and emotional processing (orbitofrontal cortex), and so on. The neo-cortex plays a major role in sleep and memory.

Structural Model of the Mind

Influenced by Darwin, Freud believed that humans are the continuation of animals and stated, "No spirits, essences, or entelechies, no superior plans or ultimate purposes are at work. The physical energies alone cause effects somehow."

According to Freud's structural model, our brain has layers of consciousness—id, ego and super ego—which are considered as three parts of the mind. Id is related to instinctive behavior, mostly controlled by the reptilian brain. Super ego is the reasoning brain, which has a moral code. Ego is the go-between the id and the super ego. The super ego can stop the id's desires. These are essentially symbolic concepts and do not represent the anatomic parts of the brain.

From birth onwards there is a longing for pleasure and an effort to avoid pain. Pleasure is the result of the satisfaction of needs, which is the id principle. However, it could be anti-social—which can create stress. However, the longing for gratification is a motivating force, which is termed as 'libido' by Freud.

Id and Instinct

Id is an uncoordinated instinctual trend. According to Freud, life instincts motivate an individual to find food and shelter. To propagate the species, id motivates the individual to have sex. These are the biological needs. Freud called these motivational instincts 'libido' (from the Latin word meaning 'I desire').

Instinct is defined as a largely inheritable and unalterable tendency of an organism to provide a complex and specific response to environmental stimuli without involving reason. It is the innate behavior of an organism, with a 'fixed action pattern' (FAP) to a definitive external stimulus. Instinctive behavior is innate or inherited, and no learning or past experience is required for its expression. Innumerable examples can be listed from nature—animal feeding, fighting, dancing of the bees, making nests by the birds, and so on. It is related to id described by Freud. Animals that do not have sufficient volitional capacity may not be able to change the FAP. However, humans can intentionally change the pattern of response to an external stimulus.

The role of instinct in behavior is inversely proportional to the complexity of the neural system. The greater the size of the cerebral cortex, the higher the role of learning (from experience and society) and lower the role of instinct. Psychologist Abraham Maslow was of the opinion that humans are devoid of instinct since they have the capacity to set aside instinctive urges.

Social learning is important for the higher forms of animals like chimpanzees, which have to learn the skill of mothering. However, many researchers believe that social learning has an instinctive basis.

For humans, the environment plays a major role in positive and negative behavior. The limbic system controls our emotions and motivation, and has a say in our instinctive behavior. These behaviors are influenced by sensory input, which is processed by the limbic system. Various evolutionary changes have taken place in the limbic system during the progression of mammals to primates. Instinctive and innate behavior is very complex and interesting as it encompasses different fields such as physiology, psychology and sociology.

Learning can counter instinct. When an animal is subjected to the same type of stimulus frequently, it becomes habituated to it and ceases to respond to the stimulus. This is an elementary form of learning.

Associative learning is one where a stimulus, which does not have any direct effect, is associated with another stimulus, which in turn produces a reaction in the animal. Ian Pavlov's theory of 'conditioned learning' explains this phenomenon. If an animal is given food after a bell is rung regularly at a particular spot, the animal will start responding to the bell by coming to the spot.

Observation and imitation are other types of learning seen in higher animals. Cultural transmission and teaching is yet another method of social learning. This method of learning is not restricted to mammals—it has been observed in lower forms as well. Ants direct their peers to the direction of food through 'tandem running', and bees do so by dancing. Mating rituals are often regarded as FAPs. Animals often fight for sexual supremacy (right to reproduce), and for supremacy in the group. Many animals, including humans, live in groups and have a symbiotic relationship with reciprocal dependence.

Id (Latin for 'it') is the instinctive brain—the animal brain or the reptilian brain. It is mainly concerned with the survival instinct (selfishness, sex and aggression) and operates on the 'pleasure principle.' Instant gratification is its motto. It is an essential ingredient for an infant's survival. Libido is a part of the id and it is the driving force behind instant gratification of desire and pleasure. Freud has said about libido: "It is filled with energy reaching it from the instincts, but it has no organization, produces no collective will, but only striking to bring about the satisfaction of the instinctual needs subject to observance of the pleasure principle." And the id "…knows no judgments, value, no good and evil, no morality."

Ego

This part of the consciousness develops at around the age of three. Id is buried underneath the ego. It is considered to be the moderating force between the id and super ego. Ego strives to keep us within the bounds of reality and social culture, and behave with common sense and reason. It is the component of reality and function in conscious, preconscious and subconscious minds. It tries to satisfy the needs of the id in a realistic and socially acceptable manner by weighing the cost-benefit ratio. Ego is organized and works on reality. Conscious awareness lives in the ego. Freud considered ego as a modified part of the id by the influence of the external world. Ego is a servant of three masters—external world, id and super ego. Ego is considered as the rider of an unruly horse—id. ('Ego' in Latin means 'I.') Because of these conflicts, three types of anxieties can arise. Objective anxiety arises from external threat, neurotic anxiety arises from the irrational gratification demands of the id, and moral anxiety comes from the threat of the super ego's demands.

Super Ego

The super ego marks maturity and is stationed in the prefrontal cortex. This area insists on morality, social consciousness and logical and reasoned behavior—resisting the gratification thrust of the id. Primates show moral behavior such as mutual assistance and cooperation. This can be considered as short-term costs for long-term gains. Morality is considered an advantageous quality for evolution. Studies have shown that people with anti-social personality disorder have lower prefrontal cortex by eleven to fifteen percent than a normal person. The super ego yearns for perfection—what can be called the conscience—and punishes the ego for wrongs with the feeling of guilt. Super ego often counters the id as the former works in a moralistic and socially acceptable way, while the id seeks instant gratification. Ego is often tormented between these two. The urge to take something from a person forcefully is anti-social. This urge is countered by the super ego and is moderated by the ego.

Topographical Model of the Mind

Sigmund Freud has proposed the theory of the mind, based on the three levels of consciousness, which is depicted as an iceberg model.

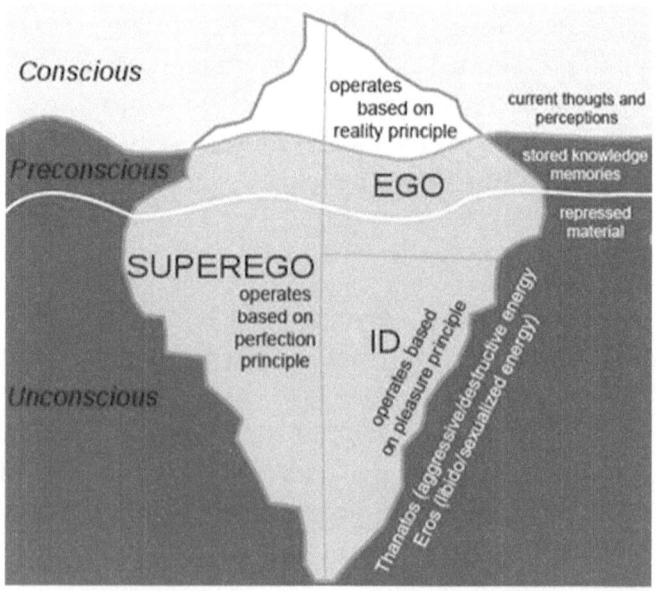

Fig 8 *Structural and topographical model of mind*

Conscious Mind

The conscious mind is composed of the mental process we routinely use while awake, and it is the rational mind. It also forms a part of the memory. This is the part of the mind you are aware of, and you can contemplate on it.

Preconscious Mind

This forms a part of the memory that can be retrieved by the conscious mind, though the person is not consciously aware of it. Thoughts, which are unconscious but not repressed, are available for cognitive processing. For example, if you want to recall where you were two days ago, you have to recall it from the preconscious mind to the conscious mind. The 'tip of the tongue' phenomenon is another common form of preconscious processing.

Unconscious Mind

This is the reservoir of feelings, thoughts, urges and memories. They may influence our behavior though we are not aware of it. Id is a part of the unconscious mind that longs for instant gratification.

Freud has related his topographical model to the structural model. The id is entirely in the unconscious part of the mind. The ego and the super ego have unconscious, preconscious and conscious parts. He has depicted the mind as an iceberg, where the major part is submerged under the water—which is the unconscious.

He also believed that this unconscious part sometimes comes to the surface through several means such as dreams, or slip of the tongue (Freudian slip). (One example Freud mentioned was a slip of the tongue by a British parliamentarian, wherein he referred to another parliamentarian who was not on good terms with him as 'the honorable member from *hell*' instead of 'Hull.' We often witness these types of slips in our daily life as well.)

The 'social science model' and 'evolutionary model' are two schools of thought regarding the development of human psychology. The former considers the mind as a blank slate at the time of birth, and the development of character is due to external influences. John B. Watson, founder of behaviorism wrote: "Give me a dozen healthy well-formed infants and my own specified world to bring them up and I will guarantee to take any one at random and train him to become any type of specialist I might select—doctor, lawyer, artist, merchant chief, and yes even a beggar—man or a thief, regardless of his talents, penchants, tendencies, abilities, vocations and race of his parents."

The latter school believes that development of the psychology of a person is mainly by inheritance, but does not reject the influence of environmental factors.

Hierarchy of Needs

Abraham Maslow proposed a theory of psychological development in 1943, which is known as the 'hierarchy of needs.'

Cognition and Confusion

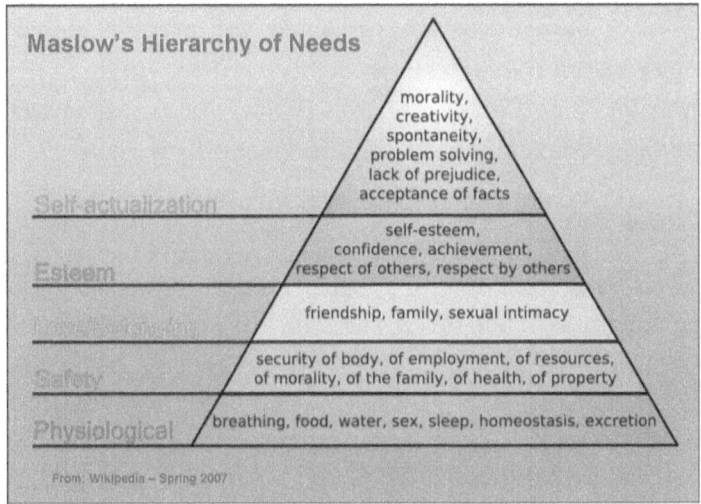

Fig 9 *Hierarchy of needs*

Physiological Needs

These are the factors needed for survival—which include metabolic needs such as air, water, food, sleep and elimination of waste. These are common for all living beings. Clothing, protection of self and sexual needs are also included in physiological needs. Semi-starvation carried out during World War II showed that the subjects became irritable and unsociable. They could think only of food; they were not concerned about their appearance or even personal hygiene. They even lost interest in sex, and they preferred pinned-up pictures of food items instead of young women. Their thoughts and dreams were about food.

Safety Needs

Once the physiological needs are satisfied, safety is the next priority. Security of job, financial security, health and wellbeing comes under this need.

Need for Love and Belonging

Need for belonging comes under interpersonal relationships such as friendship, intimacy and family bonding. There is a need to love and to be loved, both sexually and otherwise. Lack of this will lead to depression. Acceptance in the society and social groups are achieved by membership in clubs, social and professional organizations. Activities such as sports, interaction with colleagues and peers, and attending conferences are part of this need for belonging.

Esteem

This is the need for respect. Excellence in one' profession, and hobbies and social activities are certain methods used to achieve respect. It is essential that a man should have self-esteem. Imbalance in this often causes inferiority complex. There are two types of esteem:

Lower: The need for respect from others

Higher: The need for self-respect, strength, competence, independence, freedom, self-confidence and so on; lack of this results in an inferiority complex

Self-actualization

This is the final stage of the needs. Self-actualization is achieved by exploiting the full potential of a person.

In Maslow' own words, "What a man can be, he must be." Maslow believed that, to achieve self-actualization, man should master other needs.

Kurt Goldstein introduced the word 'self-actualization' as a motive for attaining one's full potential. This is the desire to become everything one is capable of becoming. According to Maslow, it is a desire. According to Goldstein, it is the 'arriving force.' Self-actualization rarely happens—say, in less than one percent of the population. The people who achieve self-actualization are people with positive qualities and attitude towards the society at large, and who have self-confidence with self-esteem.

According to Maslow, "Hierarchies are interrelated rather than sharply separated." The need for each hierarchy varies from person to person, as well as age and other factors, which control a person's mental faculty. The cultural, social and educational circumstances also play a part. Only a person who has mastered the lower needs can achieve self-actualization—the highest need.

Children have more physiological needs, adolescents have the higher need of esteem, and young adults pursue and strive for self-actualization. However, security is the need that the elderly long for.

A rough parallel of this is observed in Indian philosophy. There are four stages of life after balya (childhood) and kumara (adolescence)—(brahmacharyam or before marriage, up to the age of twenty-four); grahastham (household family life, from twenty-four to forty-eight years of age); vanaprastham (from forty-eight to seventy-two years of age); and sanyasa (spiritual life). Life has four purposes: kama (pleasure—both sex and otherwise), artha (prosperity and trying for it), dharma (performing one's duties), and moksha (enlightenment). The last mentioned is the ultimate end and objective and entails liberation from rebirth. Self-realization is when you are one with God—this is the highest purpose of life.

(According to Buddhist philosophy, the primary purpose of life is to end suffering. The desire and effort to achieve it and attachment are the primary reasons for all suffering. The only way to end suffering is to follow the path of righteousness.)

Language and Speech

"I cannot doubt that language owes its origin to the imitation and modification, aided by signs and gestures, of various natural sounds, the voices of other animals, and man's own instinctive cries."

—*Charles Darwin in* The Descent of Man, and Selection in Relation to Sex, *1871*

Pierre Paul Broca had noticed aphasia (speech deficiency) in two patients who had injuries in the posterior inferior frontal gyrus of the brain. This area was later known as 'Broca's area', which is associated with speech. Injury to this area causes deficiency in language production. However, slow-growing tumors that destroy this area need not produce aphasia, as areas nearby take over the responsibility.

Neuro-imaging studies have shown that gestures intended for communication stimulate Broca's area. Many authors have proposed that language may have

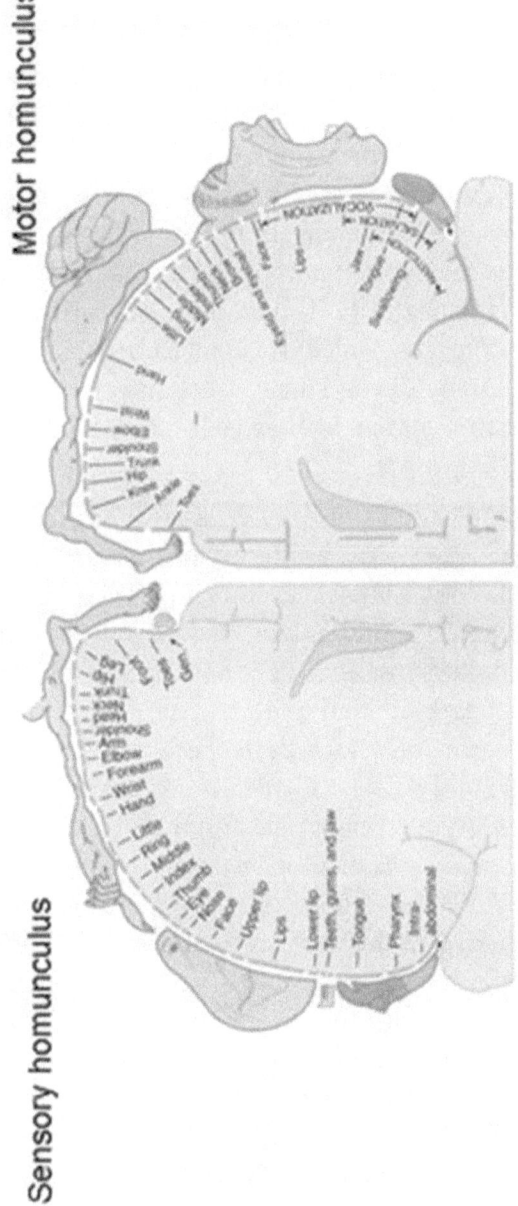

Fig 10 *The face and hands have greater representation in the brain*

evolved from gestures. Broca's area can be considered as the motor center for speech. The neural substrate in the common ancestor of apes and humans must have evolved to develop and enhance cognition and language.

Brain scans of chattering chimpanzees have shown that they use Broca's area for their chatter as well as for hearing—as in humans. In his book, *Descent of Man*, Darwin has remarked that though language is not a true instinct, as it has to be learnt, man has an instinctive tendency to speak—which is evident from the babbling of young children. Steven Pinker, a cognitive scientist, has estimated that an average six-year-old can understand about 13,000 words.

Communication by vocalization is a feature of all animals that can produce vocal sounds. It has developed to its pinnacle in humans as language and speech. Extensive studies are being conducted among Vervet monkeys to analyze their communication. It has been observed that they can produce up to ten different vocal calls, such as 'leopard call', 'snake call', 'eagle call', and so on, to which other animals respond appropriately. A pertinent question that can be raised at this point is: *Are the calls of the monkeys intentional or instinctual?* Whatever be it, the communication is taken and responded to by other Vervet monkeys. Another example worth mentioning is the worker bee—it dances when it returns with the pollen, and this dance guides the other bees as to where the pollen is. Ethnologists have done extensive studies on animal behavior. Fixed action patterns are the means by which animals of the same species communicate. The waggle dance of bees is considered as this type of communication.

Noam Chomsky, an American linguist and philosopher, was of the opinion that a single chance mutation was the cause for the development of language and speech in humans—which means that the language developed almost suddenly. This is known as the 'discontinuity theory.' However, most of the linguistic scholars believe in 'continuity theory', which proposes a gradual development of language. Proximity of the Broca's area and the motor area of the mouth and hand, and the firing of the area during gestures for communication, points to the fact that language communication has originated from gestural communication in a gradual manner. However, there is no definitive consensus among scholars.

It is interesting to note that deception in behavior has increased with the development of language. In lower forms of animals, the vocal sounds are reliable. It is less reliable among primates as they have a tendency to deceive each other. It is easy to fake words.

Deception is not restricted to higher forms. For example, the female scorpion flies expect their mates to present them with an insect. Some males pretend to be females to get the insects from prospective males.

Distraction displays are very common among birds—they mainly use this to distract the attention of the predators from the nest and their young ones. When threatened by a predator, they move away from the nest or the young ones, and pretend that they have a broken wing or are brooding. As the predator approaches, they escape from there. A broken wing is displayed by nesting waders, plovers and doves. (When I was young,

I searched, in vain, the whole bush near an abandoned pond for a wader that I thought was injured.)

There is no controversy over the fact that the ability to talk has survival value, and that it has developed by natural selection. The ability to speak and the development of language are associated with the changes in the vocal apparatus and subtle changes in the cognitive ability of the brain. Unfortunately, neither the brain nor the vocal tract was preserved in fossils. However, studies on the endocast of the skull have shown that Wernick's and Broca's area were present in the neanderthal man and in Homo habilis, which suggest that speech might have developed about two-and-a-half million years ago. (*Tobias P.V., 1981; and Holloway, 1983; Human neurobiology*)

Anatomical evolution of the organs of speech is intriguing. It is well known that monkeys and apes have specialized organs for communication by using sound. However, it is only humans who use their tongues to modulate sound for communication. The human tongue has a different shape from other primates. Because of its attachment and shape, the horizontal and vertical tubes forming the supra-laryngeal vocal tract (SVT) are almost equal in length. The movements of the jaw and tongue can change the area of the tube and alter the frequencies of sound. The vertical tract is shorter in other species.

The larynx is placed low in the neck in over this issue.

Cognition and Confusion

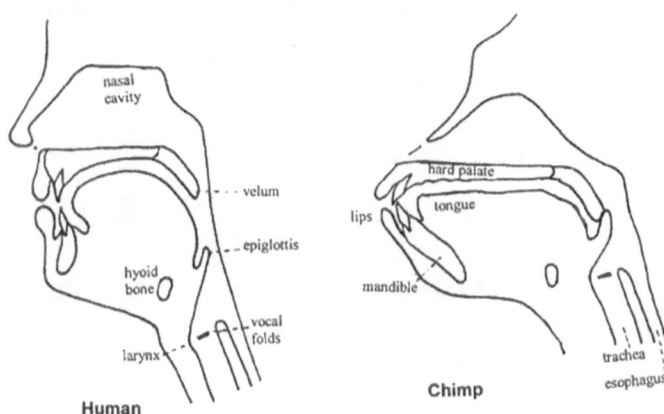

Fig 11 *Comparison of the voice box of humans and chimpanzees*

In animals, the larynx is placed high in the neck and the tongue is flat. Hence, the tongue meets the soft palate and food can pass on either side of the larynx. So, they can swallow and breathe at the same time. This is the situation for human infants as well. In adult humans, the larynx is lowly placed and the tongue is more rounded and curved downwards. There is a sound chamber behind the tongue and above the larynx, which can be adjusted to make different sounds and to shut it off from the nasal chamber.

The shape of the base of the skull may have some influence on the structure of the vocal tract. In Australopithecus and human infants, the base of the skull is flat; it became curved in Homo erectus (which lived about two million years ago). The neanderthal skull was less arched than humans. There is an argument that the shape of the skull base cannot be considered as an indicator for the shape of the vocal tract.

Though the primary function of the lip is different, it has a role to play in the production of sounds. During evolution, primates changed their lifestyle from nocturnal to diurnal; with greater reliance on vision than olfaction, they lost the wet nose. The muscles of the lip and the face have taken up the job of expression as a co-option.

(In oriental languages, the letters are grouped in a scientific manner, progressing from guttural to labial—Ka, Cha, Da, Tha and Pa, and its relatives or variations. 'Ka' is guttural; 'Cha' and its derivatives are modulated by the contact of the posterior aspect of both tongue and palate; 'Da' by the anterior part of the tongue and the anterior palate. The production of the sound 'Tha' is by the tip of the tongue and the teeth. The sound 'Pa' is still anterior and is modulated by the lips.)

The forkhead box protein P2 (FOXP2) is encoded by the FOXP2 gene located in chromosome no 7 in humans. This is a gene associated with speech and language. This gene is expressed in foetal and adult brain, heart, lung and gut. Interestingly, this gene is more active in females—which could be the reason for their superior language learning capability.

Evolution of Language

The origin of human language is not clearly understood yet—for that reason, there is no consensus among experts. Lack of fossil records of the brain and the vocal cords and the non-existence of early hominid species makes the research on the origin of the species very complex and ambiguous.

Cognition and Confusion

It is widely accepted that the development of language was a continuous process that evolved from the pre-linguistic communication methodology of the primates. Whether language was genetic, or developed by and through social interaction, is debatable.

The Broca's and Wernicke's areas in the primate brain control the muscles of the face, mouth and larynx, and also control the recognition of sounds. The regions of the cortex for movements of the mouth and hands are very close in the cortex, and the neural system for gestures for communication and vocal language are similar. So, it is assumed that human language has its origin in gestural language. Non-human primates use gestures for communication. These gestures may be genetic in nature. Each species has its own gestures—gorillas beat their chests, which some of the humans also do to show off their machismo. Vocalization in apes is related to the corresponding emotional state.

About two-and-a-half million years ago, the early Homo species emerged and bipedalism might have brought about anatomical changes in the vocal tract, making it an inverted L-shaped one that facilitated vocalization. The shape of the larynx and its low position is considered a prerequisite for many sounds, especially vowels. Since there is no definitive fossil evidence, ample controversy exists regarding the development of speech. Proto-language, which lacks syntax, was probably the beginning of language, which the early homo used.

The anatomical features that favored speech (L-shaped vocal tract) might have evolved gradually. Hence, the development of language from the early homo to the modern man might have been gradual through Homo habilis, Homo erectus and Homo heidelbergensis.

Neanderthals might have had anatomical requisites for language, but might not have developed syntactic language. Klien, who had studied stone tools extensively, argued that the neanderthals didn't have the aesthetic sense and complexity of the brain to develop a modern language. The argument is based on the fact that, after Homo habilis, stone tool technology did not progress for about two million years.

According to fossil records, anatomically modern humans emerged about 200,000 years ago in Ethiopia. However, they did not behave very differently from the earlier Homo heidelbergensis. They were probably less efficient in hunting, and they may have had to resort to agriculture because of this. Modern tools were developed around 70,000 to 50,000 years ago, and it is supposed that a fully developed language was necessary for teaching such sophistication to youngsters. *(Wolpert, L. Lewis, 2006)*.

Gene FOXP2 is associated with speech, and some consider the development of speech to be due to a sudden mutation of this gene. However, it is now known that neanderthals also shared the same gene—hence, it cannot be a mutation unique to humans.

Language could be a secondary development of cognition in humans. One of the most cognitive functions unique to humans is their capacity to ask questions. Researchers have tried to teach chimpanzees and gorillas language and to communicate by using lexigrams. They were partly successful, though their vocabulary was very limited and communication was not at all syntactic. Moreover, the apes were not able to ask questions, which is a characteristic unique to humans.

Memory

Memory involves three important processes—encoding, storage and retrieval. Encoding involves receiving the information as an external stimuli and processing it. Storage means keeping it as a record so that it can be retrieved whenever required. Memory works on the interactions of the nerve cells.

Sensory memory is an ultra-short-term memory, which lasts for a split second, decays very fast and is lost permanently. However, it is long enough to be stored in short-term memory. Humans have five basic senses, and may have five corresponding memories. The important ones are the iconic, echoic and haptic memories, which are visual, auditory and tactile respectively. Memory represents our sensory inputs, and the relevant ones are taken up to the short-term memory, which is for about ten to fifteen seconds. The important ones are transferred to long-term memory, which is kept for life.

Short-term memory is held for a short period lasting from a few seconds to one minute. The information that can be held is limited to 7 (+/-) 2 (five to nine) elements, while long-term memory can hold plenty of information.

Short-term memory, if it is being rehearsed, is taken up by the long-term memory. Some theorists believe that memory is unitary, and there cannot be any demarcation between different types of memory. Synaptic theory of short-term memory suggests that a stimulus activates a spatial pattern of activity across neurons in the brain region. The memory trace decays in time that restores neurotransmitters to the levels that existed prior to stimulus presentation. Reading aloud and individual differentiation can influence short-term memory. Diseases such as Alzheimer's can cause neuro-degeneration, adversely affecting short-term as well as long-term memory. Memory loss is a natural process that happens in the elderly.

Long-term memory can store indefinite information for a longer period of time, or even for an entire lifetime, while short-term and sensory memories can keep it for a very short period (seconds). Long-term memory is stored semantically, with logic and meaning, while short-term memory is stored acoustically. Dorsolateral prefrontal cortex of the frontal lobe and the parietal lobe are the regions concerned with short-term memory, which is formed by transient patterns of neural connections. Long-term memory, on the other hand, is formed by permanent neural connections, which are spread all over the brain. The hippocampus is an important organ for the consolidation of information in the long-term memory from the short-term one.

Sleep is important in consolidating the learned tasks in long-term memory. Neuro-imaging studies suggest that activation patterns in the sleeping brain mirror those recorded during learning, and this rehearsal is essential for consolidation. According to the Atkinson–Shiffrin model, sensory memory can be transferred to short-term memory by attention. By rehearsal, short-term memory is transferred to long-term memory, which is stored for even a lifetime and can be retrieved when required.

Working Memory

Allen Baddeley and Graham Hitch put up a model of working memory that proposes 'an active maintenance of information in the short term memory.' The concept of working memory allows explanation of our daily activities and thought processes like reasoning and learning.

Memories degrade with the passage of time. Attention plays a key role in storing information in long-term memory. Lack of attention is the main reason for absentmindedness. Areas involved in memory are the hippocampus, amygdala, striatum and mammillary bodies. The hippocampus is involved with spatial learning, and amygdala with emotional memory. The hippocampus is important in explicit memory and memory consolidation. The amygdala is responsible for memory enhancement in emotionally charged memories. However, it is not apt to demarcate memory to specific areas. Learning and memory is attributed to neuronal synapses and pathways.

Age-related memory loss is observed in the elderly. Anxiety, depression, poor nutrition, deficiency of vitamin B12, stress, alcoholism, thyroid imbalance and so on can affect the memory. Intellectual activities such as reading, writing, quizzing and psychological games, as well as physical exercise can, to some extent, prevent memory loss.

Working memory wanes as age advances. The main reasons cited are a decrease in processing speed, less capacity to hold information and lack of inhibitory control.

Memory can be compared to a jar full of water. With advancing age, there is a decline in the brain's capacity to hold the information—which means that the size of the jar is slowly reducing, as there is a gradual decrease in the number of neurons. The top layer—that is, the recent memories—flows out and is lost. The deeper layer holds the older memories and the ones that have more emotional associations, which are deeply entrenched. This is why older people remember their childhood and adolescence better, and go on talking about them. Our experiences are guided predominantly by emotions. Experiences associated with massive production of adrenaline—which increase your heart rate and blood pressure—will be remembered distinctly.

4
Conscience and Consciousness

Two remarkable things that happened after the origin of the universe (Big Bang) were the origin of life and origin of consciousness. We were able to define and understand life to some degree, although how it originated still remains a puzzle. The road map from the elements to complex organic molecules spurred the imagination of scientists to the gateway of creating life in the laboratory. Research goes on with futility, and the goal is like a wisp of grass put forth to lure the beast of burden. Theories abound, and hope has not waned but is kept burning with renewed vigor and hope.

The brain stem has many basic functions, including regulation of heart rate, breathing, sleeping, and eating. It includes the medulla oblongata (myelencephalon), pons (part of metencephalon), and midbrain (mesencephalon).

The mind has many layers, which is of evolutionary significance. The bottom layer is made up of the brainstem and reptilian brain, which controls the basic biological functions (including heart rate, breathing, eating and sleeping) and instincts imprinted in that part. The next is the limbic system, the emotional brain, which is sandwiched between the rational brain and the instinctive reptilian brain. The rational brain is stationed in the cerebral cortex that is at the top, and was the last to develop. With advancing age, the first faculty we lose is the rational brain, followed by the emotional and finally the instinctive brain and the brain stem that controls the vital functions such as the functions of the heart and breathing. When the frontal lobe of the brain, which is supposed to be the seat of intelligence, is knocked off and the rational part is affected in succession, the human is reduced to being controlled by the limbic system—and hence becomes an emotional being. Further deterioration makes the person almost an instinctive animal concerned only with food and shelter, with no concern for near and dear.

Evolution of Consciousness

The relationship between the mind and the body is a riddle. As the brain is not preserved in fossils and none of our ancestral species are alive, evolution of our brain defies elucidation. Research in neuroscience answers many questions, though not all and not completely. Analysis by experiments and observation of patients suffering from mental and brain disorder, and of the patients who suffered from brain injuries, allows only a peep into the relationship between consciousness and the brain.

Brain state and emotional state are thought to be identical, and external stimulations give rise to emotions. However, the question as to whether emotions can arise without external stimulation is a hot topic of debate among scientists and philosophers.

'Casual sense' may not agree that consciousness is the byproduct of brain activity, but rather believe that it is independent and may in fact cause brain activities. (Casual sense may find it difficult to accept that the Earth is a globe.)

The mind is the seat of many faculties, such as consciousness, emotions, thinking, perception, memory and so on. The human mind is probably the most complex object we know of—it defies definition and has not yet been deciphered properly. Whether animals have a mind—and, if so, to what level it has developed—and the relationship between the mind and brain are still incomprehensible. A very pertinent question one should ask is whether the mind has evolved by natural selection.

It is hard to define consciousness in its encompassing sense. It could be arbitrarily defined as awareness of the external world and self, sentience, the ability to experience and feel and so on. Nobody questions the existence of consciousness, but its origin, seat and its relation to the brain are a constant subject of debate, which has not yet been settled. This debate has gone on from pre-Aristotelian times to the present.

In his book, *The Selfish Gene* (1976), Richard Dawkins wrote: "The evolution of the capacity to simulate seems to have culminated in subjective consciousness. Why this should have happened is, to me, the most profound mystery facing modern biology."

Pascal opined, "The heart has its own reasons, which the head can never understand." Rousseau wrote, "Above the logic of the head is the feeling in the heart." Ricahard Frackowaik and seven other neuroscientists, in the chapter 'The Neural Correlates of Consciousness', of their book *Human Brain Function*, wrote: "We have no idea how consciousness emerges from the physical activity of the brain and we do not know whether consciousness can emerge from non-biological systems, such as computers. At this point the reader will expect to find a careful and precise definition of consciousness. You will be disappointed. Consciousness has not yet become a scientific term that can be defined in this way. Currently we all use the term consciousness in many different and often ambiguous ways. Precise definitions of different aspects of consciousness may emerge; but to make precise definitions at this stage is premature."

The main problem envisaged is the mind-body connection, which many fields of study, such as philosophy, psychology, religion, sociology, biology and neuroscience, have tried to decipher and explain. A proper explanation has not been derived yet.

The mind-body relationship is difficult to perceive. If mental events are the byproduct of the physical effects, why do emotions evolve? And if so, what survival value do they have?

According to physics, physical events are preceded, and followed, by physical effects. However, in humans, we often feel that mental events precede physical effects. Sensations, pleasant or unpleasant, may have a certain survival value (Example: Once bitten, twice shy). Though consciousness is a constant accompaniment of our mind, it remains the most mysterious.

Rene Descartes (17th century) thought that animals are automata, and humans are automata with a rational soul to control it. His dualist philosophy maintained that mind and matter are distinctly separate. He held

Conscience and Consciousness

the view that both are different kind of substances. In 1643, Princess Elizabeth of Bohemia objected to this argument, and, in a letter, wrote: "How the human soul can determine the movements of the animal spirit in the body so as to perform voluntary acts—being as it are merely a conscious substance. For the determination of movement seems always come from the moving body's being propelled—to depend on the kind of impulse it gets from what sets it in motion, or again, on the nature and shape of the latter thing's surface." The present mechanistic view is similar to the one expressed by the princess.

Another prevailing philosophy, which countered dualism, was monism, which maintained the concept of a single unifying reality. There is no consensus among monists as well. Some consider that matter, when organized in a particular way, gives rise to the mind (physicalism), while the idealists believe that thought is the real thing and matter is an illusion. Neutral monism suggests that mind and matter are two aspects of a distinct essence.

However, most philosophers with a scientific bent hold the view that the mind and body cannot be separated. In neuroscience, the mind is considered a biological process or a byproduct of the brain's activity. This concept is called 'epiphenomenalism', and considers mental phenomena as being caused by physical processes in the brain.

David Chalmers, an Australian philosopher and cognitive scientist, introduced the term 'hard problem of consciousness' and argued that the problem of experience will "persist even when the performance of all the relevant functions are explained." Understanding the process by which the physical process gives rise to subjective experience and consciousness may be the answer to this 'hard problem.'

In the article, 'Facing Up to the Problem of Consciousness', Chalmers wrote: "It is undeniable that some organisms are subjects of experience. But the question of how it is that these systems are subjects of experience is perplexing. Why is it that when our cognitive systems engage in visual and auditory information processing, we have visual or auditory experience like the quality of deep blue. How can we explain why there is something it is like to entertain a mental image, or to experience an emotion? It is widely agreed that experience arises from a physical basis, but we have no good explanation of why and how it so arises. Why should physical processing give rise to a rich inner life at all? It seems objectively unreasonable that it should, and yet it does."

Thomas Henry Huxley remarked, "How it is that any thing so remarkable as a state of consciousness comes about as the result of irritating nervous tissue, is just as unaccountable as the appearance of the Djin when Aladdin rubbed his lamp." (*The Elements of Physiology and Hygiene: A Text-book for Educational Institutions* by T.H. Huxley & W.J. Youmans, 1868, pp 178)

Philosophers like David Chalmers and Alfred North Whitehead are of the opinion that conscious experience is a fundamental constituent of the universe, a form of 'panpsychism.'

Panpsychism is an old theory attributed to philosophers like Thales, Plato and Spinoza, and seen in eastern philosophies such as Vedanta and Mahayana Buddhism. It is the concept that consciousness, mind and soul encompass all the things including living and non-living. The recent rise of the 'hard problem of consciousness' has aroused interest in Panpsychism.

(Vedanta is an eastern philosophy that seeks the relationship between atman [soul], brahman [god], and the world. There are three schools of thought in Vedanta. According to Advaita Vedanta, there is no difference between atman and brahman. According to Dvaita, jivatman is totally different from brahman. In Vishisht Advaita, jivatman is a part of brahman, and similar but not identical.)

Some philosophers, such as Daniel Dennet (American philosopher and cognitive scientist), feel that the 'hard problem' will cease once we understand what consciousness is, and this will be explained as a natural phenomenon.

Peter Hecker, a British philosopher, considered the present state of consciousness study as a waste of time. He sarcastically stated, "The whole endeavor of the consciousness studies community is absurd—they are in pursuit of chimera. They misunderstand the nature of consciousness. The conception of consciousness, which they have, is incoherent. The questions they are asking don't make sense. They have to go back to the drawing board and start all over again." (Chimera is a fire-breathing female monster in Greek philosophy. It is said to have a lion's head, a goat's body, and a serpent's tail.)

Mathematician-turned-philosopher Alfred North Whitehead, who co-authored *Principia Mathematica* with his student, Bertrand Russell, was of the opinion that reality consists of events rather than matter. He argued, "There is urgency in coming to see the world as a web of interrelated process of which we are integral parts, so that all our choices and action have consequences of the world around us." He visualized consciousness as an intrinsic property of creation and argued that it goes all the way down to the particles.

The brain is superbly complex, especially due to the presence of complex neuronal connections. A set of neuronal connections gives rise to neuronal events, which beget a particular conscious perception. The minimal set required is the 'neuronal correlates of consciousness' (NCC). This is basically the neurobiological approach, and considers consciousness as a property of complex neurological process. Advances in technology, which empower the neuroscientists to manipulate the neuronal correlates of neurons in animals as well as in humans, may solve the yet inexplicable mystery of consciousness.

As the prevailing controversy about the relationship between brain activity and consciousness continues, there is no consensus regarding the evolution of consciousness. Whether consciousness has any survival value is also a hot topic of discussion among scientists and philosophers.

It is debatable as to which stage of the evolutionary ladder consciousness could have emerged. That consciousness emerged along with the development of nervous system is one argument. The opposite extreme is that consciousness emerged as a unique quality in the first humans.

T.H. Huxley defended the 'epiphenomena theory' (which holds the view that consciousness is an inert effect of neural activity). William James, an American philosopher and psychologist, put forth an evolutionary argument in in his essay, 'Are We Automata', that consciousness emerged as a result of 'natural selection', which means that it has certain survival value. Sir Karl Raimund Popper (philosopher and professor at the London School of Economics), in his book, *The Self and Its Brain* puts forth the same argument.

If we accept the fact that consciousness is a byproduct of the brain's activity, a pertinent question arises: Were the lower forms of animals also conscious? When Darwin proposed the theory of consciousness, he maintained the view that evolution is purposeless, and that it just happens by natural selection. Many did not accept this premise, though they agreed with the theory of natural selection.

Ethology is the study of animal behavior under natural conditions. It is not possible to understand what goes on in the mind of an animal—hence, it is not possible to clearly decipher their level of consciousness. The level of consciousness for lower forms of animals will be very different from that of humans. Though they do not have a communicative language, they communicate each other for survival. The most important function of any living being is reproduction—for sustenance of the species and betterment, if possible. It appears that they do not have self-awareness to the extent that humans do.

Many of the higher mammals, especially the domesticated ones, show emotions such as affection, hatred, sorrow and excitement. They even dream. To say that they are not conscious is difficult to accept. Most of these lower forms may not be self-aware. They may not recognize themselves in a mirror. Birds are known to fight with their own images reflected in windowpanes. Many of them have superior senses than us. Dogs can hear and smell better than us. The loss of olfactory reception centers during human evolution could explain our rudimentary sense of smell. Development of color vision in humans might have moderated the need for smell. Owls can see well in the night; many fishes can detect electricity and use them for navigation; birds have a well-developed navigation system that helps them to even navigate to distant continents. It is difficult to decipher at which level consciousness could have evolved. Single-celled organisms may have the lowest levels of awareness. As multicellular organisms evolved, they developed many more senses and sensory organs that were transferred to the brain—which was also enlarged by the addition of new areas. In primates, it was the neo-cortex, with the development of new abilities along with the enhancement of consciousness. Donald Griffin, an American professor of Zoology who did an extensive study on animal behavior in *The Question of Animal Behavior* (1976), argued that animals are conscious, just like humans.

Humans' well-developed communication system—language—can be attributed to the evolution of the voice box. The development of language catapulted the extra genomal and extra neuronal knowledge, and paved the way for cultural evolution. Collective knowledge could be transferred through generations. Written language enhanced the knowledge bank. With the development of the digital world and computers, knowledge is at our fingertips. (Excessive reliance on computers is adversely affecting the intellectual faculty of humans—many have to resort to calculators for even simple calculations.)

Introspection is another important attribute that the human mind developed—in other words, we could talk to ourselves and we could reason. We could use past experiences to project future outcomes. We developed the capacity to plan, predict and execute things. During the evolutionary process, our thought process progressed from logically linear to intuitively complex.

As we cannot communicate with other animals in their language, we are unable to understand their feelings and emotions. However, it is very clear that as per the theory of 'survival of the fittest' that all existing species are fit for survival. Since their vision and other senses are totally different from ours, their impression of the world will be totally different from ours—and could be even more realistic, as they merge with nature to a greater degree than us.

Eastern philosophy accords a soul (atman) to every living creature. Scientifically, as well, we cannot deny the existence of consciousness in the lowest of lowly animals.

The statement 'God sleeps in minerals, awakens in plants, walks in animals and thinks in humans' by Arthur Young is a profound one that evokes contemplation. Consciousness might have evolved very slowly from lower animals to higher animals and to its present, highest, form in humans. No perfect definition on consciousness has been arrived upon with consensus.

Could 'atman' in Indian philosophy be translated to 'consciousness'? The atman is present in all the living creatures. After the death of an organism, it is reborn as another life form. Depending on the way the organism lived (good or bad), the next birth is decided. If it was good, the next birth is as a higher being. The cycle of death and rebirth continues until the atman reaches nirvana—the end of rebirths and merging with brahman or universal consciousness, which is equivalent to God. According to Eastern philosophy, the ultimate aim of life is perfection.

A subtle parallel can be drawn scientifically. According to the concept of rebirth, it is the atman that is transferred to the new being, and there is no death for the atman. In a living organism, the only thing that does not have death is the genome in the gamete, which pairs with the gamete of the sexual partner and is transferred to the offspring. So, the offspring can be considered as the rebirth, and the gamete can be considered as atman. (According to Indian dualist philosophy, there should be union of 'purusha' and 'prakrithi' for the emergence of 'budhi' or intellect). Further, it is a good and ethical life that helps the atman be reborn as a superior being. The 'good life' means better genes. In short, favorable changes that take place in the genome are transferred to the offspring (rebirth). Accumulated favorable changes beget new superior species—which is the crux of evolution.

Evolution of Emotions

Different emotions might have developed at different times, and are related to the evolution of brain. Primary emotions, which are instinctive like fear, are essential for protection of life. This, along with the early part of the brain, evolved in forms that came before the mammals. Maternal affection towards the offspring and kin seems to have evolved in mammals. Guilt, pride, confidence and so on are social emotions that developed in primates.

Affection, anger, angst, anguish, annoyance, anxiety, apathy, arousal, awe, boredom, confidence, contempt, contentment, courage, curiosity, depression, desire, despair, disappointment, disgust, distrust, dread, ecstasy, embarrassment, envy, euphoria, excitement, fear, frustration, gratitude, grief, guilt, happiness, hatred, hope, horror, hostility, hurt, hysteria, indifference, interest, jealousy, joy, loathing, loneliness, love, lust, outrage, panic, passion, pity, pleasure, pride, rage, regret, relief, remorse, sadness, satisfaction, schadenfreude, self-confidence, shame, shock, shyness, sorrow, suffering, surprise, terror, trust, wonder, worry, zeal and zest (Wikipedia) are common emotions experienced by humans and by some animals—these emotions have a profound bearing on self-awareness and consciousness.

In 1872, Charles Darwin wrote *The Expression of the Emotion in Man and Animals*, in which he proposed that emotion has an evolutionary history. He studied not only facial expressions in humans and animals, but also tried to find parallels in their behavior. He also observed that similarities in the expression of emotions are more among related species.

The first of three theories he proposed was 'principle of serviceable habits'—useful habits inherited by the offspring. One example he used is of the lowering of the eyebrows, which helps to block the light entering the eye, and raising the eyebrows to allow for more light and better visibility. Lowered eyebrows express displeasure. Some people raise their eyebrows while trying to remember something, or to express surprise. His second principle is 'antithesis' such as shrugging the shoulders, which does not serve any purpose and is the opposite of confidence. The third principle is expressive habits—a nervous discharge. Expressions shown during extreme situations such as anger, fear or pain are certain examples.

Anger, fear, disgust, sadness, and enjoyment are emotions that have maximum universal similarity, according to the research conducted by Paul Ekman.

Joseph LeDoux has conducted research on fear. Fear can be evoked by two systems. In the first one, sensory information goes to the thalamus and then to the amygdala, and the autonomic and motor reaction (fear) is elicited. This is the fast track. In the second one (which is a more recent evolution), the information from the thalamus goes to the appropriate sensory cortex (visual to the visual cortex, and touch to the somatosensory cortex) and from there to the frontal association areas. The gauging occurs in the frontal area that communicates with the amygdala directly and modifies the response (increase or decrease). The older, fast track system helps instant reaction and it could save lives.

Emotions are controlled by the limbic system (in Latin, 'limbus' means 'border'). These parts lie above the brain stem and below the higher center—the cerebral cortex. This borders the lower part, which is the animal part of the brain (which controls instincts and autonomic control of the body), and the higher part or the cerebral cortex (which controls the reasoning and planning).

Intuition

"The only real valuable thing is intuition."

— **Albert Einstein**

"When you reach the end of what you should know, you will be at the beginning of what you should sense."

— **Khalil Gibran**

Intuition is a hunch, a feeling that has no rational basis. It is a strong feeling that one cannot explain. It originates from our inner self. Intuition is not properly understood and belongs to no-man's land and, hence, is discussed in science, philosophy and religion. Intuition is often misunderstood as instinct, meaning, truth and so on. Sri Aurobindo considered intuition to be in the higher realms of intelligence, and thought that, as the mind evolves, its main function will be intuition. Osho also considered it almost the same way. Advaitha, a popular and renowned philosophical thought in India, considers intuition as an experience to realize brahman.

Intuition is a mental process by which past experiences and knowledge are quickly processed and weighed against the situation presented to you. It is a quick processing mechanism. Intuition is more accurate for people in situations that they have experienced before. Studies on brain-damaged people who have lost their intuitive capability showed that they took lot of time to differentiate between items of a similar nature, as they spent a lot of time assessing it logically. Intuition helps people take decisions very quickly. For example, many senior, experienced doctors can intuitively judge the ailment of the patient accurately as soon as he walks into his office; less experienced doctors may not be right in judging a patient. The more experienced a person, the better the chances for his/her intuition to be correct. Intuition is a distilled product of reason, knowledge and experience. This could be the reason why intuitions from great minds are considered as revelations. However, to rely completely on your hunch may sometimes be disastrous.

The modern approach of psychology to intuition is different. Carl Jung defined intuition as 'perception via the unconscious', for which sensory perception is the starting point and is refined by the unconscious faculties. Intuition is different from instinct and it is probably a refined product of sensory perception and past experiences and observations stored in the memory. Intuition could be the next level after intelligence.

Evolution of Spirituality

Spirituality emanates from the concept that there is an ultimate purpose for life. Life intends to reach certain higher realms of existence after the physical body ceases to exist. It is a higher level of evolution (apart from biological evolution). (*Piyasīlo*-1991)

Conscience and Consciousness

William Sturgis Bigelow, a Buddhist physician, endeavored to unify the concept of biological evolution by natural selection to spiritual evolution in his book, *Buddhism and Immortality*. He argued that when an individual emerges from 'unconditioned consciousness' and ascends by natural selection, he/she reaches a level of celestial experience.

Samkhya, a system of Hindu philosophy (about 2,500 years old), which is atheistic and dualistic, conceptualized the union of purusha (pure spirit) and prakrithi (psychophysical nature) as disturbing the equilibrium and descending to the physical world. The purpose of evolution is the release of purusha and its return to spiritual quiescence.

Alfred Russel Wallace, co-discoverer of evolution by natural selection, in his book, *World of Life* (1911), described evolution as a "creative power, directive mind and ultimate purpose." He thought that it was not possible to explain intelligence or morality by natural selection and attributed its origin to "the unseen universe of spirit."

Charles Darwin believed that the humans were evolving into higher realms, which is evident from the following statement in his book, *The Descent of Man*: "Man may be excused for feeling some pride at having risen, though not through his own exertions, to the very summit of the organic scale; and the fact of his having thus risen, instead of having aboriginally placed there, may give him hopes for a still higher destiny in the distant future."

Sri Aurobindo was an Indian nationalist, philosopher and yogi, guru and poet. In his books, *The Synthesis of Yoga* and *Life Divine*, he explains his concept of 'spiritual evolution.' He argued that matter has an impulse for life; that life evolved from matter, and mind from life. He believed that evolution has a purpose. He described progression from inanimate matter to a future state of divine consciousness, which he called 'super mind.'

There were many other renowned people who thought that there is a greater purpose for evolution. Another important person was Teilhard de Chardin, a Jesuit paleontologist, who had an important role in discovering the 'Peking Man.' He presented a teleological view of the universe—cosmic evolution, from atoms to molecules, life and man—that is, progression from biosphere to 'noosphere' (sphere of human thought). According to him, evolution will continue on to the Omega point, which he identifies as Jesus Christ.

Carl Gustav Jung, a famous psychiatrist, believed that life has spiritual goals beyond material achievements. He studied most of the religions, and the journey to this higher realm—which he called 'individuation'—which is at the mystical heart of all religions.

Towards Perfection

Perfection is flawless. We attribute perfection to God. The question raised here is: Are we on our path to perfection through evolution? It is not an easy task to answer this question. There are arguments for and against it. I have already put forth some of the opinions of great men. Humans have never experienced perfection, though

Conscience and Consciousness

we often use the word 'perfection' for 'excellence.' So it can be considered as a relative and not absolute term. Perfection can be put as 'omnipresent, omniscient and omnipotent'—the attributes of God. According to Christian philosophy, humans who live a good life become beings that are next to God. For Christians, God is personified. According to Hindu philosophy, humans have to go up the ladder through various rebirths and finally merge with the universal consciousness, which is akin to God itself. From this standpoint, it is very clear that awareness has come to us along with a longing for power, knowledge and reach. This yearning may be the reason as to why humans created God as his ultimate wish. Most religions have created God in man's own image. According to Jean-Baptiste Lamarck, There is an innate nature, or 'inner want' in every living being to be modified, which he called as a 'complexing force.' Unfortunately, science is not able to test it, as it is an exacting principle cannot accept that cannot be tested physically. The evolutionary trend is towards modifications for adapting to the environment. During the early days of biological evolution, it was passive existence, and survival and reproduction were the prime (or only) resolutions. Hence, adaptation to the environment was the tactic. As the evolution progressed, life forms started modifying the environment to their needs. As awareness and consciousness evolved, the thirst for knowledge and power overtook the life process. All these phenomena point to the premise that life is progressing towards perfection, where all the positive qualities like function, knowledge and beauty meet.

Frederick Nietzsche stated: "Equality and democracy are against the grain of selection and survival, that not masses but geniuses are the goal of evolution, that not justice but power is the arbiter of all differences and all destinies."

The second statement of Lamarck (1744–1829), who proposed the theory of evolution, was that every living organism, including humans, has a tendency to reach a greater level of perfection. Charles Darwin was influenced by Lamarck's theories.

T.H. Huxley, Darwin's friend, stated, "It is an error to imagine that evolution signifies a constant tendency to increased perfection. That process undoubtedly involves a constant remodeling of the organism in adaptation to new conditions; but it depends on the nature of those conditions whether the direction of the modifications effected shall be upward or downward." Thomas Huxley's statement clearly shows his stand as a strong believer of the evolutionary process, which has no specific purpose, but as a random adaptive process by natural selection.

However, he was perturbed by the justification of the unethical behavior of the strong against the weak. So, he stated later. "Men in society are undoubtedly subject to the cosmic process...Social progress means a checking of the cosmic process at every step and the substitution for it of another, which may be called the ethical process; the end of which is not the survival of those who may happen to be the fittest but of those who are ethically the best."

The conflict between the survival of the fittest by competition, which amounts to oppression of the weak and the cultural values he believed in, might have pained and confused him.

Whitehead's idea of God, which he called 'The brief Galilean vision of humility', is as follows: "It does not emphasize the ruling Caesar, or the ruthless moralist, or the unmoved mover. It dwells upon the tender elements in the world, which slowly and in quietness operates by love; and it finds purpose in the present immediacy of a kingdom not of this world. Love neither rules, nor is it unmoved; also it is a little oblivious as to morals. It does not look to the future; for it finds its own reward in the immediate present."

Mutation is one of the major processes in natural selection, and it is random. Natural selection, as the name denotes, selects the favorable ones for the survival. Many scientists are vehemently against the argument that natural selection has a tendency towards perfection, and feel that the argument stems from the unconscious presumption that life is an entity separate from nature.

The 'Gaia hypothesis' was proposed by James Lovelock and co-developed by microbiologist Lynn Margulis in the 1970s. This hypothesis proposes that the interaction between organic life forms and the inorganic environment forms a 'self-regulating complex system', which maintains the conditions suitable for living on this planet. Lovelock considered the planet as a huge organism. He was of the opinion that biological feedback mechanism could evolve by natural selection, and those organisms that can improve the environment had a better prospect for survival. Though this could not stand the acid test of science, the concept has positive points—it gives importance to nature as an all-encompassing mother, which sustains life. There is an interrelationship between all the living and non-living in the planet.

A homeostasis is maintained through the interactions of biosphere, atmosphere, oceans and the land.

The 'Neo Darwinian' theory of evolution is based on the concept of natural selection through the survival of the fittest by competition. According to Lynn Margulis (a microbiologist who had done extensive research on microorganisms), the impetus of evolution is not competition but cooperation. This theory is known as 'Symbiogenesis.'

Theosophy does not contradict evolution but considers it as a divine plan, and that the Supreme Being permeates all living and non-living. Natural selection often has certain evolutionary advantages and features—it has to be inheritable, and should have functional advantages that help to increase fitness. However, in nature, we see some adaptations have no specific function but it may later acquire some function. For example, the feathers of the birds are essential for flight. The feathers probably might have developed as insulation for the two-legged dinosaurs called theropods—the closest relatives of birds that had feathers but could not fly. A more acceptable argument may be that the feathers on the forelimb might have helped the dinosaurs to run faster—otherwise, there is no reason for it to be selected. However, there are certain characteristics that might be a chance occurrence or a byproduct of another characteristic. Some characteristics might have developed for a different purpose and then become coopted for a different function. Evolution is also driven by factors such as migration, mutation, and genetic drift.

In 1988, Richard Laski of the Michigan State University in East Lansing started studying the changes that have taken place in E. coli bacteria (considered to be a workhorse in genetic studies) for generations. He studied them for more than 500,000 generations and observed that they go on evolving, and postulated that there is no upper limit for evolution and that fitness never stopped increasing.

As humans evolved from the early primates, importance has been shifted from physical to intellectual and cognitive abilities. If natural evolution is allowed to progress, Homo sapiens may evolve to a more beautiful and intelligent being—'Homo sapiens divine'—which I have shown in the graph projecting the facial proportions to the future. In about three-and-a-half million years, the changes that have taken place from the Australopithecus (Lucy) to the modern man are mind-boggling. Hence, we can envisage the changes that can happen in another few million years. Not just interplanetary, but interstellar travel may become routine. In short, he may become the master of the universe, which is almost equal to being God. However, unfortunately our intelligence has devised eugenics already, and has started developing artificial intelligence. In the near future, we will create not only artificial life and creatures, but also develop life-extending techniques, which may stall our natural evolution. Thus, it may not be Homo sapiens divine but biological robots that will be ruling the world in the future.

5
The Ladder/Tree of Evolution

Organic evolution is the progression of organisms through time. After the Big Bang, elements transformed from a plasma state into elements and molecules. Carbon, water, nitrogen and phosphorous were the most important ones needed for life to originate. From simple molecules, complex compounds that had the capacity for duplication were formed. The most important ones for life per se were amino acids, proteins and nucleic acids. Chemical evolution gave rise to organic life. Subsequently, organic evolution resulted in varied organisms—from single-celled prokaryotes and eukaryotes to complex plants, animals and birds.

An important piece of evidence for common origin is the cell structure, which is very similar in all living organisms—be it plants or animals. Differences between cells of varied organisms are too little when compared to the physiological and biochemical similarities.

Charles Darwin described a logical and scientifically acceptable principle and explanation for evolution in his book, *Origin of Species*. One objection raised by creationists against evolution is that it is not observed in daily life. The reason it cannot be observed is because millions of years are required for the development of a new species. Fossil records accord important evidence that can support the theory of evolution. One of the important fossil records is the evolution of humans. We shared a common ancestor with the chimpanzees, and then we branched off from them around three-and-a-half million years ago. The species, Australopithecus afarensis (named Lucy) is supposed to be our great-great-grandmother. From then on, through around seven to eight different species (some of which lived at the same time), we have evolved to the modern man, Homo sapien (the term means 'wise human'). We do have the fossil records for all these different species. By reviewing the body frames and skulls of these ancestors, we can indubitably appreciate the transition of the primate through the early hominids and archaic man to the modern man.

Survival is different in the offspring. Each individual carries different characteristics—the ones that carry adaptive characteristics have a better chance of survival. Sexual reproduction was a quantum jump in evolution—it provided a chance for the mixing of genes of the partners during fertilization, and for the offspring to show varied characteristics. The ones that adapted better thrived, and the ones that adapted less perished. Sexuality developed very early in the evolution of life forms. Simple organisms were able to exchange or mix their genes with other organisms of the same level. However, in higher organisms, this was not possible. If, by chance, a mix does takes place between these organisms, it will be lethal for the offspring or the offspring will be sterile. (Example: A horse has sixty-four chromosomes and a donkey has sixty-two. A mule is the offspring of a male donkey and a female horse. So, the mule has sixty-three chromosomes, which is an odd number. Variations in the structure and number of the chromosome prevent pairing during meiosis. Hence, the formation of embryos does not take place and they are infertile.)

The environment had a great role to play in the process of evolution. It took around 2,000 million years for life to start on our planet after its birth as a fireball. There were mass extinctions on several occasions. *(Example: The extinction of the dinosaurs happened about sixty-five million years ago, either due to the impact of a large asteroid or comet or due to multiple volcano eruptions that produced dust affecting sunlight and subsequent photosynthesis. This reduced the available number of plants, thus resulting in scarcity of food for large, vegetarian dinosaurs.)*

The term 'survival of the fittest' is often misunderstood. If that was literally the case, there was no reason for the evolution of multicellular organisms. The larger the organism, the more challenging is its survival. The so-called most evolved of living beings, Homo sapiens, is probably the only species that has to toil hard to earn its food. None of the animals sow, cultivate, reap and store food. *(In the book,* Genesis of Bible, *God curses man to live by the sweat of his forehead.)* Moreover, man is probably the least adapted to the environment. He is always in strife with the surroundings and environment.

Evidence for Organic Evolution

Structural similarities between related species are one of the important features of biologic evolution. Evidences for evolution can be directly understood from fossil records (paleontological evidence), comparative anatomy, embryology, physiology, biochemistry and genetics.

Fossil Records

The word 'fossil' is derived from a Latin word, 'fossilium', which means 'something dug up.' The bones and teeth of animals, dug up from stratum layers of the Earth's crust are the fossils. These could be the remnants of certain existing species or of extinct ones. There are different ways by which fossilization takes place.

One of the main processes by which fossilization takes place is by petrification, wherein the hard parts and (rarely) the soft parts are replaced by minerals such as iron pyrites, calcium carbonate and silica. Occasionally, the entire specimen may be frozen in extreme cold situations. In Siberia, an extinct mammoth that was about 25,000 years old was found—the body was so well preserved that the dogs ate it. Footprints and trails in soft mud could harden to form fossils. Casts and molds are another form of fossilization. The formation of fossils is very rare as the conditions for fossilization are very stringent—the body should have been buried immediately before decay, and that too in favorable sites such as water-borne sedimentary rocks.

The dating of fossils became almost accurate after using radioactive isotopes for calculating the half-life. The fossil's age is estimated by calculating the decay of the element after the death of the organism. Carbon 14 (C^{14}) has a half-life of 5,730 years, hence anything older than 25,000 years (five half-lives) cannot be calculated. Uranium dating can fix the date for a fairly long duration, as the half-life of uranium is 4,500,000,000 years. Uranium decays to become lead. Recently, potassium 40 (K^{40}) is being used—it decays into argon, rubidium and strontium. The half-life of K^{40} is 1.3 billion years. Since potassium concentration in rocks is more accurate, it is better for fixing the age than uranium.

Geologically, six eras are being considered:

Archaeozoic (period of primitive life—4,000 to 2,500 million years ago; oxygen-free atmosphere; life started around 3,500 million years ago)

Proterozoic (period of early life—2,500 to 541 million years ago; pre-Cambrian evolution of soft bodied multicellular organisms)

Paleozoic (period of ancient life—541 to 252.17 million years ago; Cambrian evolution—fish, arthropods, amphibians and reptiles evolved)

Mesozoic (Period of medieval life—252.17 to sixty-six million years ago; age of reptiles)

Cenozoic (period of modern life—sixty-six million years ago; age of mammals)

During these eras, a great many changes had taken place in the environment, which resulted in dramatic changes in some life forms and extinction of many. These changes were an impetus for the evolution of new life forms. There are about five major extinction events in the history of earth (Jack Sepkoski, David M. Raup; 1982). They are the Cretaceous–Paleocene extinction event around sixty-six million years ago; Triassic–Jurassic Event around 200 million years ago; Permian–Triassic extinction event around 251 million years ago;

Late Devonian event around 375 to 360 million years ago; and Ordovician–Silurian around 450 to 440 million years ago.

During each event, about seventy to seventy-five percent of the existing species became extinct. Mass extinctions are considered to have promoted evolution of life forms. For example, the end of the cretaceous mass extinction eliminated most of the non-avian dinosaurs and made way for the dominance of mammals. Opinions vary regarding the cause of mass extinctions—drastic ecological changes, large comets or asteroids hitting earth and so on are considered to be the reasons for mass extinctions. According to Arens. N.C and West. I.D (2008), large extinctions could be due to short-term shock due to long-term stress on the biosphere.

Fig 12   *Evolution of elephants*

The Ladder/Tree of Evolution

Fossil records unequivocally prove the concept of evolution, since fossils appear in different periods. None of the past forms simulate the present ones. The primitive ones are usually seen in deeper layers of the Earth's crust and the more evolved, complex ones are seen higher up in the strata.

Transitional forms showing serial evolution are available from fossil records. A very popular transitional form is the Berlin specimen of Archaeopteryx lithografica (150 to 145 million years ago)—a transitional form from the dinosaur to the bird. This fossil shows the characteristics of a primitive bird and a theropod, and is one of the first transitional forms seen in the fossil records. This suggests that birds evolved from reptiles.

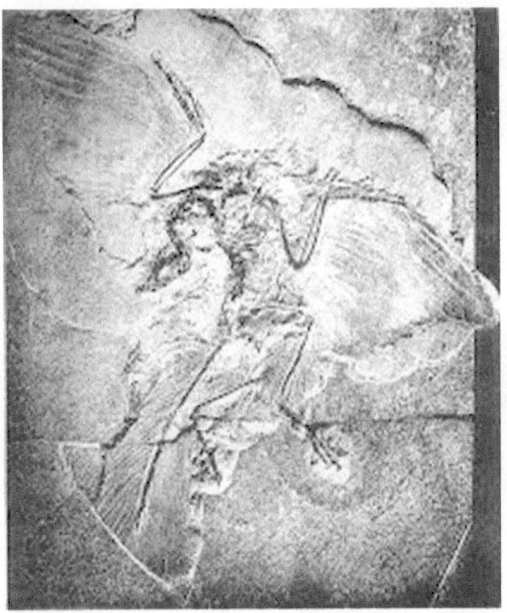

Fig 13   *Archaeopteryx lithografica—fossil*

Geographical Distribution

Geological changes and the drifting of landmasses created certain isolated areas and islands that were not easily accessible. In such regions, independent variations took place. The lack of competition also created variations. A striking example observed by Darwin is in the Galapagos archipelago, about 500 to 600 miles off the coast of Ecuador. He found giant tortoises and two species of iguana lizards, which are not found anywhere else. He found twenty-six species of birds, out of which twenty-five were unique to the archipelago.

Comparative anatomy

Carl von Linne (1707–1778) was a pioneer in classifying the living organisms, and this branch is known as taxonomy. Grouping organisms according to their morphological characteristics reveals their evolutionary relationship. A typical example is the skeletal structure of mammals, including whales and bats. They have similar structures, handed down from a common ancestor that was modified for different purposes. These structures are called 'homologous structures', since they have descended from a similar ancestor. Most mammals, including aquatic whales, bats, arboreal monkeys, terrestrial cat and man, have almost the same number of bones, arranged in a similar fashion, though they are serving different functions. This is considered to be strong evidence for evolution.

The Ladder/Tree of Evolution

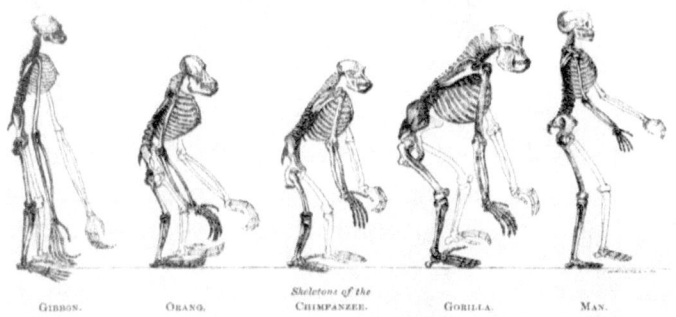

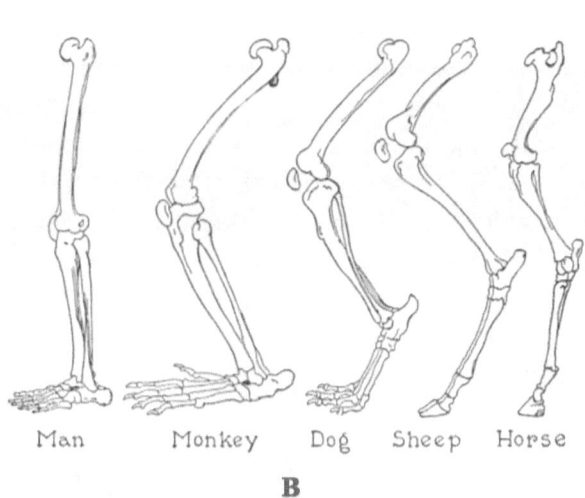

Fig 14 *(A) Comparison of skeletons of primates (B) Skeletal structure of mammals including man and horse—they have similar structures, known as homologous structures*

It has to be noted that organisms from totally different groups may also have similar structures because they developed in a similar environment.

These are called 'analogous structures.' Analogous structures with structural resemblance are known as 'homoplastic.' 'Convergent' evolution is the development of similar adaptations seen in certain unrelated organisms that live in the same habitat. An example is the fusiform-like body for aquatic mammals such as dolphins and fish. This shows the capacity of the species to cope with the environment.

Vestigial organs—structures that don't have any perceivable function—seen in many species are near clear-cut evidence for evolution. This occurs as a part of normal evolutionary process, due to the changing environment. Humans have around a hundred vestigial organs. The important ones are vermiform appendix, coccyx (fused tail vertebrae), muscles that move the ear, nictitating membrane of the eye, wisdom teeth and so on. Vestigial organs are often functionless, but were functional in their ancestors. (Though vermiform appendix has no digestive function, it may have some immunological functions.) Another example is that of pythons, which have rudimentary hind limbs. Whales have vestiges of pelvic girdle (ilium) and femur.

In 1798, Etienne Geoffrey Saint-Hilaire (1772–1844) stated: "Nature never works by rapid jumps, and she always leaves vestiges of an organ (even though it is completely superfluous), if that organ plays an important role in other species of the same family."

Comparative Embryology

Embryology is the study of the development of an organism from a fertilized egg to its natural adult form.

All complex and multicellular organisms start their life from a fertilized egg known as the zygote. The zygote multiplies and forms a blastula, and differentiates to two fundamental germ layers—ectoderm and endoderm. A third germ layer—mesoderm—develops later. Up to a certain stage, embryological development is very similar for different organisms, and then deviates to different paths to maturation. The similarity in development is more and extended in related species. This is evident from a comparison of embryological development in different species.

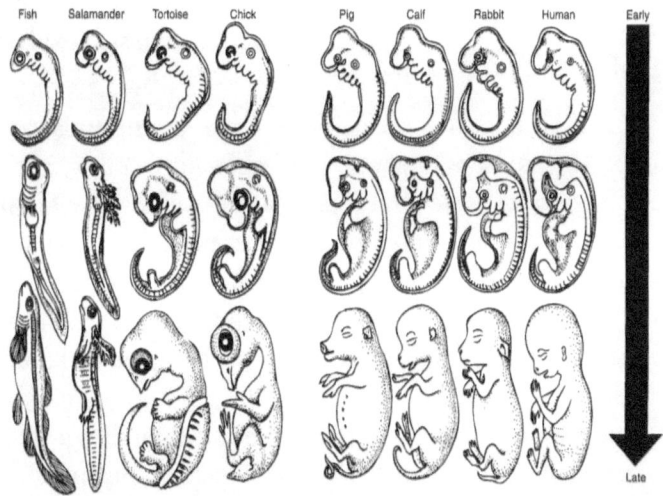

Fig 15 *Comparative embryology—up to a certain stage, embryological development is very similar for different organisms and then deviates to different paths to maturation*

The embryos of complex and advanced animals repeat many stages of development present in lower animals. Ernst Haeckel (1834–1919), a strong supporter of evolution, studied the embryological development of vertebrates and proposed his famous principle of 'biogenetic law', which says 'ontogeny recapitulate phylogeny.' ('Ontogeny' means embryological development, and 'phylogeny' means evolutionary history.) His idea gave the impression that every organism is repeating all the stages of evolution it has gone through. Many biologists refuted this idea, and considered it an oversimplification. Many believed that the developmental sequence is only an evidence for their common ancestry. Von Baer (1792–1876), who said that embryos of higher animals repeat many stages of development seen in lower animals, proposed the concept of 'recapitulation.' This principle is widely better accepted than the Haekelian concept.

During the early stages, embryos of all animals look alike. All have pharyngeal pouches and pharyngeal clefts. Haekel argued that this resembled the gills of fish, and also proposed that this represented an adult fish's development. All develop limb buds and embryonic tails. For example, a mammalian heart goes through the stages of the hearts of fishes and reptiles. The Haekelian concept is outmoded; scientists confirm Darwin's view of 'embryonic stages are similar among related species.' However, the growth of embryos reflects evolution.

The embryological development of mammals is considered as evidence of evolution as it goes through the stages of a fish, amphibian and reptile. During the evolutionary process, they have accumulated many characteristics through mutations, but they retained the genes of the early stages of development. During embryological development, early stages are influenced by the genes of fish, amphibians and reptiles, in the order of evolution respectively, and later by mammalian genes. In short, it can be said that characteristics of phyla, classes and genera prop up during early stages of development and finally characteristics of the species develop.

(Drosophila melanogaster, the common fruit fly, is a workhorse in genetic studies. The genome sequence of this fly was published in 2000. About seventy-five percent of known human disease genes have matches in the genome of fruit fly. In spite of such a gap in evolution between humans and arthropods, many fundamentals of genetics of Drosophila melanogaster are relevant to humans.)

Chromosome

Physiology, biochemistry and metabolism of all the living organisms are very similar. The chromosomes and its arrangement are in the same fashion. This suggests a common origin for all organisms. Similarities in the enzymes, hormones, and antigen antibody reactions are all glaring examples of a common origin for all life forms.

In all living beings, twenty amino acids are arranged in different sequences in different numbers and patterns to form different varieties of proteins. Genes control the production of these proteins. There are just four types of nucleotides—adenine, cytosine, guanine and thymine in the DNA. Adenine pairs with thymine and guanine with cytosine. These nucleotides are arranged like winding stair connected by weak links in the chromosome, which is present in all cells of all organisms. Three pairs of nucleotides form a codon. In total, there are only sixty-four types of codons (four times four times four). Each codon is responsible for particular amino acid. A series of codons together forms a gene, which is responsible for the production of proteins. These features are common for all living organisms.

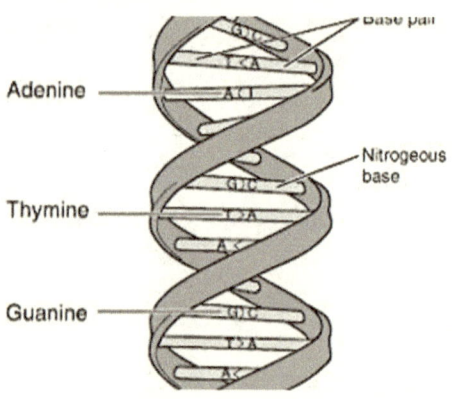

Fig 16   *Double helix structure of DNA*

By comparing the sequences of nucleotides in a DNA strand, an evolutionary pathway can be deciphered. The change of sequences in the codons helps in assessing the likely family tree and the relationship of the organisms. It is interesting to note that humans share 98.4 percent genes with the chimpanzee, 97.7 percent genes with the gorilla and 96.4 percent with the Orangutan. Chimpanzees have twenty-four pairs of chromosomes and humans have twenty-three pairs. Chromosomes 12 and 13 (later named as 2A and 2B) of the chimpanzee fused to form chromosome number 2 of humans.

The antiserum of the antigen of an animal can be tested with antigens of another animal to assess their relationship. Serum proteins of a chimpanzee are more similar to humans than that of apes, gorilla or baboons.

Comparing the amino acid chain of hemoglobin of shark and humans, we can conclude that humans and sharks diverged around 460 million years ago.

Theories and Mechanisms of Evolution

'Unity in diversity' is a characteristic that can be used meaningfully about life on earth. The number of species that ever lived on this planet is anybody's guess—which ranges from thirty-five to 4,000 billion—of which 99.9 percent are extinct through great extinction episodes. For complex organisms, the average life span of the genus is about four million years. We humans are almost there (B. Bryson). From a virus to the most complex human, morphologically, and in many other characteristics, diversity is striking. But when we go deeper and compare the basic structure of the cells, the pattern of the chromosome, metabolism, reproduction and so on, we find a striking unity between all living beings.

Charles Darwin and Alfred Russell Wallace were the first biologists to give a tangible argument about the evolution of the organism from simple to complex forms. Biologists generally agreed that *'descent with modification'* is the mechanism for progression of life. Diversity of the biological process is the hallmark of evolution.

Evolutionary thought was present from pre-Socratic Greek philosophers such as Anaximander and Empedocies. They thought that one type of animal could descend from another type. Aristotle (384–322 BC), one of the earliest philosophers, classified all living beings in hierarchal order in his *'Scala Natura'* or *'great chain of being.'* Here, he placed plants at the bottom and humans at the top. The natural phenomenon of evolution was not accepted, even in the 17th century. The biological classification by Carl Linne, a Swedish botanist, in 1735, viewed the creation of species as a part of the divine plan. Pierre Louis Maupertuis (1698–1759), a French mathematician and philosopher, wrote that modification happening during reproduction, which was accumulated through many generations, was responsible for the development of new species. His observations are considered as a precursor of the theory of 'natural selection.'

'Temple of Nature', a poem written by Erasmus Darwin (1731–1802), Charles Darwin's grandfather, which was published posthumously, contains his belief in evolution of modern life from simple and minute organisms. Comte de Buffon and many others had their own theories about evolution.

Jean-Baptiste Lamarck (1744–1829), a French naturalist, was an early proponent of 'evolution through natural phenomenon.' Lamarck put forward two main postulates. The first was that changes in animals were influenced by the environment. He sited examples such as blindness in moles, the presence of teeth in mammals and lack of teeth in birds, and rudimentary vestiges due to loss of function. The second proposal was that life was structured in an orderly manner such that each part is synchronized for the organic movements of animals. In other words, his theory had two parts—

new structures appear because of the organism's 'inner want', and structures are acquired in response to need and environmental pressures. These structures are inherited—or, acquired characteristics are inherited. The postulate of 'inner want' cannot be tested scientifically and the theory of 'inheritance of acquired characteristics' was proved wrong. Hence, the Lamarckian theory of evolution was not favored. Somatic cells do not pass on characteristics to germ cells.

Lamarck believed that there is a tendency for all living organism to become more complex. He referred to this as a 'complexing force' (le pouvoir de la vie)—a force that tends to be 'in order.' He took the principles of alchemy to explain this principle. Adaptation of an organism to the environment is carried over to the next generation and, gradually, the progress of an organism to new species takes place. Otherwise, the use and disuse of an organ is preserved by reproduction for the new individual. This particular concept called 'soft inheritance' was initially accepted but has been rejected by modern genetics science.

However, in epigenetics, there is proof that soft inheritance plays a part in the changing of some organism's phenotypes. For instance, methylation, an epigenetic change, can prevent DNA expression. So, modern molecular biology is now forced to reconsider Lamarckian concept.

Though the Lamarckian concept of 'inheritance of acquired characteristic' was rejected by the scientific community, he is considered to have the drawn the attention of the scientific world towards evolution. Charles Darwin himself accepted the role of use and disuse as subsidiary mechanism in natural selection. (More than 116 plant species and about 103 marine species and many animal species were named in honor of Lamarck.)

Charles Robert Darwin (1809–1882) was one of great biologists that ever lived and contributed to the theory of evolution, which has stood the test of time. (Darwin discontinued medical studies as the subjects and lectures didn't interest him. His father, a wealthy medical practitioner, wanted him to become an Anglican parson, but he was more interested in natural science.) He proposed that all species descended from a common ancestor. With Alfred Russel Wallace, he published the scientific theory that 'natural selection resulting from struggle for existence' is the mechanism for branching of organisms into different species. (In 1858, Alfred Russel Wallace independently developed this idea, and wrote an essay on this and sent it to Darwin.)

Charles Darwin's work can be attributed to his voyage on the *HMS Beagle* with Captain Robert Fitzroy, where he was more a companion than a naturalist collector. The ship left Plymouth on 27 December, 1831. The *Beagle*'s voyage lasted five years. They visited many islands in the Atlantic Ocean, the South coast of South America and some islands of South Pacific—of which the Galapagos Islands were the most important. Darwin saw nature as a dynamic force, constantly achieving and retaining equilibrium. He spent most of his time on land for natural history collections and investigating geology. He kept detailed notes about everything and the specimens were sent to Cambridge with detailed descriptions.

In 1859, he published his famous book, *On the Origin of Species*, with indisputable evidence for his theory of evolution. This is considered as the most important book of the nineteenth century. His five years' experience gained aboard the *HMS Beagle* established him as an eminent geologist and author. Intrigued by the geographical distribution of wild life and their variations he noticed among related life forms due to isolation, especially in the Galapagos Islands, he conducted investigations and conceived the idea of natural selection in 1838. Darwin's work established 'descent with modification' as the primary cause for diversification and evolution. In 1871, he wrote *Descent of Man, and Selection in Relation to Sex*, where he examined human evolution and sexual selection and proposed that men and apes have a common ancestor. His other book on humans is *The Expression of the Emotions in Man and Animals*.

(When Darwin published *On the Origin of Species* in 1859, the wife of the Bishop of Worcester was so distressed that she said, "Let us hope that it is not true, but, if it is, let us pray that it does not become generally known.")

Thanks to his publications, intuitive nature, hard work and association with important scientists, Darwin became a prominent figure in the scientific community. During his research, he found time for reading as well. His reading included 'An Essay on the Principle of Population' by Thomas Robert Malthus. In this essay, Malthus asserted that population, when unchecked, would double in twenty-five years, that it grows by geometrical progression and would exceed the food supply—known as the 'Malthusian catastrophe.' Darwin wrote in his autobiography: "In October 1938, that is fifteen months after I started my systematic enquiry, I happened to read for amusement, 'Malthus on population', and well prepared to appreciate the struggle for existence, which goes on everywhere, from long continued observation of the habits of animals and plants. It at once struck me that under these circumstances favorable variations tend to be preserved, and unfavorable ones to be destroyed. The result of this would be the formation of new species. Here then I have got theory by which to work." Selective breeding practiced by farmers, which yielded different varieties, also influenced him.

(Wallace was also inspired by Malthusian theory, and came to a similar theory as that of Darwin and wrote to Darwin—which was almost like a bombshell for him. Darwin sent this paper to Charles Lyell and offered to keep quiet about his theory and his work. However, C. Lyell and Joseph Dalton Hooker, who knew about his work and his theory, persuaded Darwin, and a joint paper was sent to the Linnean Society. This paper was presented in the society by Lyell and Hooker but did not create much of an impact in the scientific world. However, the publication of *On the Origin of Species* on 24 November, 1859, took the scientific world and community by storm. In 1864, Darwin was awarded the Copley Medal—the greatest honor of the Royal Society.)

The essence of Darwin's theory is: "The change in species by the survival of an organism, exhibiting a natural variation that gives it an adaptive advantage in an environment, thus leading to a new environmental equilibrium, is evolution by natural selection." Natural

The Ladder/Tree of Evolution

selection is a continuous process by trial and error and rejection and selection.

Any pair of animals, birds or plants produces plenty of offspring, and, if all survive, that will be a geometrical progression, which is impossible for the Earth to sustain. Naturally, only the fitter ones survive and weaker ones are eliminated. For example, an oyster lays eighty million eggs in its lifetime. If all the eggs mature and reproduce, the Earth would be filled with oysters. There is a natural struggle for existence and only the healthy ones will survive to keep the population of oysters almost constant. There would be a struggle for existence within the species and between different species for food and space, which are supposed to be the biological needs for survival. Moreover, there will be adaptations, powered by mutation, to the pressure from environment, and favorable adaptations are carried on to the next generation. This is the basic concept of 'natural selection' and 'survival of the fittest.' Characteristics of the surviving ones will be transmitted to their offspring—and, through each generation, there is a betterment of the species. When a species, through environmental and other factors like migration, gets isolated, there will be varied adaptive changes that will be transmitted to the next generations. Through time, new species may form.

Darwin's theory can be summarized in two statements from the introduction and conclusion of *Origin of Species*:

"As many more individuals than can possibly survive; and as, consequently, there is a frequently recurring struggle for existence, it follows that any being, it vary however slightly in any manner profitable to itself, under the complex and varying conditions of life, will have a better chance of surviving, and thus be naturally selected. From the strong principle of inheritance, any selected variety will tend to propagate its new and modified form."

And,

"There is grandeur in this view of life, with its several powers, having been breathed into a few forms or into one; and that, while this planet has gone cycling on according to the fixed law of gravity, from so simple a beginning endless forms most beautiful and most wonderful have been, and are being evolved."

Though Darwin proposed natural selection as the main driving force of evolution, he did not rule out the Lamarckian theory of inheritance of acquired characteristics. Darwin called this hypothesis 'pangenesis' and explained it in his book, *Variation in Plants and Animals Under Domestication*. He believed that environmental pressures such as use and disuse could affect the germ cells and get inherited. Wallace was opposed to this view.

Philosopher Herbert Spencer and German anatomist Ernst Haeckel saw evolution as an inherently progressive process, and the Lamarckian theory appealed to them as well. Edward Drinker Cope (1840–1897), Alpheus Spring Packard (1798–1884) and Herbert Spencer (1820–1903) tried to modify Lamarck's theory. They considered adaptation as universal, and that it arises as a result of a casual relationship of structure, function and environment. This modification was known as Neo Lamarckism. George Bernard Shaw (1856–1950) and Arthur Koestler (1905–1983) were supporters of Neo Lamarckism. Many believed that Darwinian theory makes the organism a puppet of the environment and is a mechanistic theory, while Lamarckism allowed the individual to decide its own destiny. (Herbert Spencer coined the phrase 'survival of the fittest.')

There is only a small difference between these theories, in the sense that selection always prefers one that adapts to the surroundings or environment, which is an inherent property encrypted in the genes

The Darwin-Wallace theory had certain shortcomings—it could not explain the development of new organs or over-specializations such as the mammoths' spiral tusk of mammoths, the huge antlers of the Irish deer, the huge dinosaurs of the Mesozoic period, and so on. Moreover, natural selection could not explain the degeneration of organs. Darwin would have been well aware of the limitations and shortcomings of his theories. This could be the reason for his embrace the Lamarckian concept of 'use and disuse' as 'pangenesis' in 1868. He also introduced sexual selection as a secondary driving force in evolution, to explain over-specialization or the development of ornate secondary sexual characteristics, which has nothing much to do with survival, except for display, to scare combatants and to attract mates

Darwin's book, *The Descent of Man and Selection in Relation to Sex* (1871), set out evidence that humans are actually animals that show continuity of physical and mental attributes. He presented sexual selection to explain impractical animal features such as the plumage of the peacock. In his book, *The expression of emotions in man and animals* (1872), he discussed the evolution of human psychology and its continuity with animal behavior. He concluded it with the statement, *"Man with all his noble qualities with sympathy which he feels for the most debased, with benevolence which extends not only to other men but to the humblest living creature, with his god like intellect which has penetrated into the movements and constitution of the solar system—with all these exalted powers—man still bears in his bodily frame the indelible stamp of his lowly origin."*

Fig 17   *Charles Darwin*

Fig 18   *Alfred Russel Wallace*

In 1879, Darwin wrote, *"I have never been an atheist in the sense of denying the existence of a God—I think that generally an agnostic would be the most correct description of my state of mind."*

(At this point, it has to be noted that Gregor Mendel (1822–1884), an Augustinian monk, was doing experiments on pea plants to study the ways of inheritance of certain characteristics. He published his paper in 1866, but it was noticed only after three decades. If it had been noticed earlier, the inheritance of adaptive changes and natural selection of Darwin-Wallace theory of natural selection could have been better explained.)

The theory of natural selection had some real difficulties, which defied a proper explanation— such as the development of the eyes, ossicles of the ear, instincts, and many other specialized organs that need a multitude of small variations, and the ways by which these were selected to achieve such specialized organs. The eyes are so complex that, to develop a fully functional eye, each step should have been profitable. Light-sensitive pigments found in the eyes of vertebrates and invertebrates are closely related. The evolution of the eye was a complicated issue because it involved multifarious factors, each of which would have been beneficial to the individual for natural selection. The eye is an example of an analogous organ present in several taxa developed independently. It is presumed that the complex eye developed during the Cambrian evolutionary explosion (about 540 million years ago).

The eyes have developed differently in different organisms to meet their requirements. The eye's properties in various organisms vary in visual acuity, perception of light wavelength, sensitivity to intensity of light, detection of motion and discrimination of color. The design of the eye is a miracle, and Charles Darwin himself considered the evolution of the eye as a riddle. However, the feasibility of its development would have been possible only if each step was beneficial for the organism.

Whether the eyes developed independently in different taxa, or had a common origin is debatable. However, the protein responsible for light sensitivity is opsin and its subfamilies. These proteins were present in the last common ancestor of all animals.

Mutations occur in about one in four thousand gametes. Studies by eminent scientists such as R.A. Fischer (1890–1962), J.B.S. Haldane (1892–1964) and others have found that a favorable characteristic acquired by mutation takes around 560 generations for half the population to have this gene. It is about fifteen years for Drosophila melanogaster (the common fruit fly), 280 years for rabbits and 14,000 years for humans *(An Anatomy of Thought* by Ian Glynn). A pertinent question that lurks in our minds is whether four billion years from the time of the birth of our planet was sufficient for life to form and evolve to this present form.

Altruism is also not easily explained by natural selection. However, Richard Dawkins has tried an explanation for altruism in his book, *Selfish Gene*. He proposed that competition is actually between genes, and individual organisms are just short-lived vehicles for this competition.

Endo Symbiotic Theory

The theme of this theory, proposed by Lynn Margulis (1938–2011), is that interdependence and cooperative existence of multiple prokaryotic organisms evolved over millions of years into eukariotic cells. This theory is widely accepted as the important mechanism for the rise of organelles. Margulis formulated her theory in 1966 and her book, *Origin of Eukaryotic Cell*, discusses the concept in detail. Her theory gained wider support when it was found that the genetic material of mitochondria and chloroplast are different from the DNA of the host cell. She formulated another theory—that the symbiotic relationships between different organisms of different phyla and kingdoms are the driving force in evolution.

Gaia Hypothesis

Also known as 'Gaia theory' (1971), this hypothesis was proposed by James Lovelock and co-developed by Lynn Margulis. The proposal is that the Earth is a self-regulating system, and co-evolved with the interaction between organic and inorganic world. Evidence used includes the development of an oxygen-rich atmosphere, which supports complex life from thermo-acido-philic and methanogenic bacteria. The activity of photosynthetic bacteria during the pre-Cambrian period transformed Earth into a planet with an oxygen-rich aerobic atmosphere, thus helping the evolution of more complex eukaryotic life.

Timeline of Evolution

About one billion years after the origin of life, prokaryotes such as cyanobacteria, with the capability for photosynthesis appeared and gave rise to large quantities of oxygen on earth. By about two-and-a-half billion years ago, anaerobic organisms were almost wiped out and aerobic organisms that utilized oxygen for their metabolism emerged. This is known as the 'great oxygenation event.' This is probably the time that life started emerging on land. After another 400 million years, complex cells that had a nucleus, called eukaryotes, appeared. The split between bacteria and archaea occurred during this time. By around 1.2 billion years back, during the Proterozoic eon, sexual reproduction evolved, which lead to faster evolution. Nine hundred million years ago, choanoflagellates emerged—they are supposed to be the ancestors of all animals. They lived in colonies and have showed primitive cellular specialization. The earliest multicellular animal is said to be a sponge that had a very simple form, and partially differentiated tissues. About 550 million years ago, bilateria (animals with front and back, and bilateral symmetry) appeared. Fish and proto-amphibians appeared in due course. Plants appeared around 475 million years ago and reptiles by 300 million years ago. Mammals and birds appeared between 200 and 160 million years ago. Primates appeared about sixty million years ago. Great apes walked on the earth about twenty million years ago. Human predecessors belonging to the genus Homo appeared around two-and-a-half million years ago. Anatomically, modern humans emerged around 200,000 years ago.

In its 4.6 billion years of circling the Sun, the Earth has harbored an increasing diversity of life forms:

The Ladder/Tree of Evolution

3.6 billion years ago—simple cells (prokaryotes)

3.4 billion years ago—cyanobacteria (performing photosynthesis)

2 billion years ago—complex cells (eukaryotes)

1.2 billion years ago—eukaryotes (which sexually reproduce)

1 billion years ago—multicellular life

600 million years ago—simple animals

550 million years ago—bilaterians (water life forms with a front and a back)

500 million years ago—fish and proto-amphibians

475 million years ago—land plants

400 million years ago—insects and seeds

360 million years ago—amphibians

300 million years ago—reptiles

200 million years ago—mammals

150 million years ago—birds

130 million years ago—flowers

60 million years ago—the primates

20 million years ago—the family Hominidae (great apes)

2.5 million years ago—the genus Homo (including humans and their predecessors)

200,000 years ago—anatomically modern humans

Periodic extinctions have temporarily reduced diversity:

2.4 billion years ago—many obligate anaerobes became extinct in the oxygen catastrophe

252 million years ago—the trilobites became extinct in the Permian–Triassic extinction event

66 million years ago—the pterosaurs and non-avian dinosaurs became extinct in the Cretaceous–Paleogene extinction event

(Dates are approximate)

Major Extinction Events

It is estimated that ninety-eight percent of the documented species are now extinct. From 540 million years ago to the present, several extinction episodes occurred. Hence, this is considered a phanerozoic phenomenon (belonging to the current eon). There were five major episodes of extinction and several minor ones. The major ones are listed below.

1) Ordovician–Silurian Extinction

    This occurred around 450 to 440 million years ago. Seven percent of all families, fifty-seven percent of all genera, and sixty to seventy percent of all species went extinct. This was the second largest of all extinctions.

2) Late Devonian Extinction

    This has happened about 370 to 360 million years ago. Ninety percent of all families, fifty-seven percent of all genera, sixty to seventy percent of all species got extinct, and this event lasted for twenty million years. The causes are not very clear—global cooling, oceanic volcanism and so on have been considered as some reasons.

The Ladder/Tree of Evolution

3) Permian–Triassic Extinction

   This happened around 251 million years ago. This was the largest extinction event ever. Fifty-four percent of families, eighty-three percent of genera, ninety to ninety-three percent of all species (fifty-three percent of marine families and eighty-four percent of marine genera, ninety-six percent of marine species and seventy percent of land species) became extinct. This ended the primacy of reptiles. There was also a mass extinction of insects. This extinction is also known as the 'great dying'; recovery took more than ten million years.

4) Triassic–Jurassic Extinction Event

   This marks the boundary of Triassic and Jurassic periods. After this extinction, dinosaurs gained dominance on land. This happened about 200 million years ago. During this extinction, twenty-three percent of all families, forty-eight percent of all genera, seventy to seventy-five percent of all species became extinct. Most of the non-dinosaurian archosaurs, most therapsids and most large amphibians became extinct. Gradual climatic change, sea level fluctuation, acidification and so on could be the reasons for this extinction

5) Cretaceous–Paleogene Extinction Event.

   This is also referred to as the end-Cretaceous/ KT extinction or K-Pg extinction. About seventeen percent of all families, fifty percent of all genera and seventy-five percent of all species became extinct (Raup D.; Sepkoski J.J.; 1993). This extinction event, which happened about sixty-five million years ago, eliminated all non-avian dinosaurs, which had been the dominant animals on land. With this extinction, mammals and birds emerged as the dominant land vertebrates. This extinction was due to the massive impact of a comet or an asteroid. This resulted in an extended winter, which compromised photosynthesis (Alvarez L.W.). A layer of Iridium, abundant in asteroids, and not found much in earth's crust, at the boundary layer, is evidence of astral impact.

Importance of Extinctions in Evolution

Extinctions ended the dominance of certain groups on the planet. The dominance of dinosaurs changed to dominance of mammals during the KT extinction, while previous extinctions had helped the dinosaurs to gain dominance. It is thought that extinctions have helped or promoted evolution. Charles Darwin was of the opinion that struggle for existence was the most important factor in promoting evolution. Several reasons have been cited for extinction events—flood basalt event, falling of the sea level, impact of asteroids, global cooling/warming, anoxic events, methane release, oceanic overturn, gamma ray burst, plate tectonics and so on. Recovery after extinctions took around five to ten million years; in severe cases, it took fifteen to thirty million years.

Genetics

Genetics is the science of heredity. There are two types of heredity variations—recombination and mutation.

Recombination is by hybridization—when new combinations arise. By mutation, new genetic material is produced. Point mutations produced many varieties in both plants and animals, and such accumulated variations later resulted in speciation. Variation in the number of chromosomes is another reason for speciation.

Genetics deals with the transmission of characteristics from parent to offspring—through inheritance or heredity. Heredity and variation is an important aspect of speciation. Gregor Mendel (1822–1884), through his experiments on pea plants in 1866, founded the science of genetics. Mendel's study remained obscure for more than three decades, until Hugo de Vries and others rediscovered it in 1900. (William Bateson, a proponent of Mendel's work, coined the word 'genetics' in 1905. It was derived from the Greek word 'genesis' or 'origin.') Subsequently, there was tremendous research and progress in this field.

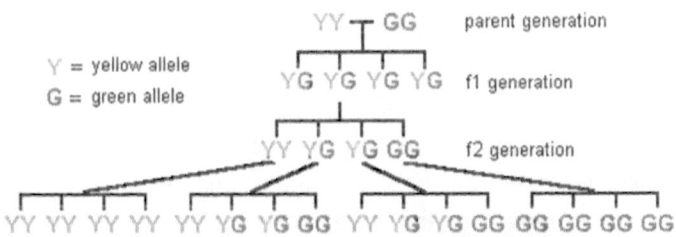

Fig 19 *When two homozygous parents with different alleles breed, the first filial generation will have genes from each parent. The dominant allele will be expressed. This chart shows the way the genes are distributed*

Mendel conducted breeding experiments on garden peas that had different characteristics in terms of the height, color of the flower, and color of the seed. His findings helped Carl Correns, one of the rediscovers of Mendel's ideas, to formulate three laws.

Fig 20   *Gregor Mendel*

1) Law of dominance
2) Law of segregation
3) Law of independent assortment

Mendel postulated that genes exist in pairs, and that the dominant one is expressed. When a tall pure-bred pea plant is crossed with a dwarf pure-bred one, the first-generation plant will be tall even though they have the genes for both characteristics, since the gene for better height is dominant.

This can be expressed as:

TT × tt = Tt, Tt, Tt, Tt. 1st filial generation (all plants are tall)

Tt × Tt = TT, Tt, tT, tt. 2nd filial generation (three plants are tall and one plant is a dwarf)

The Ladder/Tree of Evolution

With the discovery of chromosomes, genes and cell division (mitosis and meiosis), the understanding of genetics has enhanced and influenced all fields of biology, including the science of evolution.

James D. Watson and Francis Crick discovered the structure of DNA in 1953 (they were awarded the Nobel Prize for their discovery). Inheritance can be explained from the structure of DNA. The DNA acts as a template to produce messenger RNA, with molecules very similar to DNA, and create an amino acid sequence to produce proteins. Each codon has three nucleotide base pairs (adenine pairs with thymine, and cytosine pairs with guanine in DNA; in RNA, thymine is replaced by uracil). Each codon produces one of the twenty amino acids, though there are sixty-four different types of codons from nucleotide combinations. A set of codons is responsible for the production of proteins. This set is the 'gene.' Since genes appear in pairs (one from each sexual partner), the dominant one is expressed. During meiosis there will be recombination of genes by exchange of sections between chromosome pairs. This is an important factor in the variation in offspring and natural selection.

Formation of Gametes: Meiosis is also known as reduction division by which gametes are produced. Chromosomes are seen as pairs in the normal cell (one set from each parent). During reduction division, the chromosome exchange sections and then divide to two cells with the cell having only one set (half the number) this is known as the haploid cell which is the gamete.

The important discoveries in genetics are:

1875: Strausberger discovered chromosomes

1902: Sutton discovered genes

1911: Wilhelm Johannsen formulated the concept of genotype and phenotype

1934: T.H. Morgan contributed to the nature of the genes and received the Nobel prize

1946: H.J. Muller Induced mutation in dorsophila (fruit fly) by X-rays and received the Nobel prize for the same

1958: G.B. Beadle and J. Lederbug received the Nobel prize for their contribution to biochemical aspects

1959: A. Kornberg and S. Ochoa received the Nobel Prize for the chemistry of DNA and RNA

1962: Watson, Crick and Wilkins received the Nobel Prize for the discovery of the double-helix structure of DNA

1965: Jacob, Monod, Lwoff, discovered the class of genes that regulate other genes

1966: Nirenberg and Khorana discovered the DNA code (genetic code) and received the Nobel prize

1968: First restriction enzyme described

1972: First recombinant DNA produced—joining DNA molecules from different species and inserting it in a host cell (a bacterium)

1973: First animal gene is cloned (the gene of an African frog was transferred to E. coli and the specific frog protein was produced)

1975: DNA sequencing was achieved by two groups of researchers separately—Frederic Sanger and colleagues; Alan Maxam and Walter Gilbert

1976: First genetic engineering company by Herbert Boyer produced the first human protein (human insulin in bacterium) and marketed it in 1982

1981: First transgenic mice and fruit flies

1982: Genetic bank database formed

1983: Polymerase chain reaction (PCR) invented—it could make billions of copies of a specific segment of DNA

1986: Positional cloning of a disease gene

1987: First comprehensive genetic map

1987: Artificial chromosome made from yeast could carry large fragments of human DNA—this made the mapping of the human genome easier

1990: Launch of human genome project (about 3.2 billion letters of the human genome)

1996: Mouse genetic map completed—mice share almost all their genes with humans, and about eighty-five percent of the genes are identical

1999: Chromosome 22 of the humans was the first one to be sequenced

2000: Human genome working draft was completed

2003: Human genome sequencing completed (99.9 percent accurate)

2004: Vaccine for human papilloma virus invented

2007: Traced the evolutionary origin of HIV by analyzing its genetic mutations

2008: The first gene therapy for cancer, utilizing a form of adenovirus to carry a replacement gene coding for the p53 protein

2009: Elizabeth H. Blackburn, Carol W. Greiderd and Jack W. Szostak received the Nobel prize for 'How the chromosomes are protected by telomeres and the enzyme telomerase'

2012: Sir John B. Gurdon and Shynia Yamanaka received the Nobel prize for the discovery that mature cells can be reprogrammed to become pluripotent

(DNA sequencing is a technology by which the sequence of nucleotides in DNA can be determined. In 1977, Frederick Sanger and team developed a technique of 'chain-termination sequencing', which is still used for sequencing. This technique is beneficial for studying the molecular sequence of plenty of human diseases. It is also used to find the complete sequences of human genome [human genome project] that was completed in 2003.)

Neo Darwinism

Neo Darwinism is a synthesis of Mendelian genetics with the Darwinian concept of natural selection and other biological branches. The concept of 'pangenesis'—the Lamarckian concept supported by Darwin—is removed from this. Now, the widely accepted theory is Neo-Darwinism, often called 'Modern Evolutionary Synthesis.' One of the main issues is the difficulty in explaining the macro-evolution observed by paleontologists.

The rediscovery of Gregor Mendel's laws of heredity by Carl Correns and Hugo de Vries in 1900 allowed for

The Ladder/Tree of Evolution

explanation of natural selection and speciation. The contribution of each parent retained its integrity, and the factors given to offspring by parents were segregated and not mixed. Hugo de Vries (1886) put forward the theory of mutation—where there is a sudden change. He proposed that mutation and not selection is the primary factor in evolution.

Mutation happens when changes take place in the nucleotide sequence of the genome of an organism during meiosis and multiplication of the DNA. There can be several ways by which mutation can occur—duplication of sections of DNA, genetic recombination, duplication of gene, point mutation (change of one nucleotide). Changes in number of chromosomes may cause larger mutations. This happens when large sections of the chromosome break away and rearrange to form a new chromosome. In hominids, two chromosomes fused together to form human chromosome number 2. These accelerate divergence and prevent inter-breeding.

(A question to be answered is that the change in number of chromosomes cannot happen together in a group, but will be happening in a single individual and that individual will not be able to produce an offspring as fertilization cannot take place between individuals having different number of chromosomes and if at all this happens the offspring will be sterile. Parthenogenesis is an exception, but it doesn't happen in higher mammals.)

Transposons are mobile genetic material (sequences of DNA) that play an important role in the evolution of genomes. This genetic material can move about in the genome and substitute the genes and beget genetic diversity. Beneficial mutations are selected.

The gene pool is the sum total of the alleles in a population. A gene pool can be altered by mutation and hybridization, or by natural selection. Spatial and geographical isolation helped the formation of new species over time. During meiotic division, recombination takes place and new phenotypes may emerge with survival advantages or disadvantages. Advantages are naturally selected, and they will affect the gene pool. It also has to be understood that a single gene does not always control a characteristic, but by multiple 'regulatory genes.'

In short, we can understand that the processes underlying evolution are the following—mutation, changes in chromosome number and structure, recombination, natural selection and reproductive isolation. Natural selection is powered by adaptive changes, and migration helps reproductive isolation. Hybridization between races or closely related species increases genetic variability.

The nature of speciation is of different types. The one that is observed from fossils is called 'phyletic evolution.' This is evolution in a lineage, which is a modification over time. Over a long period of time, due to accumulated changes in a particular population, the present species may be totally different from its distant ancestor. Otherwise, species 'A' might have evolved to species 'B', and 'A' might have become extinct

However, this cannot be the major method of speciation. Migration and isolation could be an important method of multiple speciations. Mutation is a universal genetic phenomenon, but most mutations are not good for the individual. Occasionally, some mutations may be beneficial. These mutations increase the heterogeneity of the population. This does not amount to speciation in higher animals. However, in lower forms, which propagate by asexual reproduction, spontaneous speciation may happen due to mutations, polyploidy, fission or fusion of the chromosomes.

Instantaneous mutation via macro-genesis or saltation is questionable. Saltation evolution is a leap from the existing gene pool, resulting in a 'hopeful monster' as described by Richard Goldschmidt (1878–1958) in 1940. It is highly improbable as a change in a single individual can affect a speciation, as it is difficult to find a mate to continue the lineage. In animals that reproduce parthenogenetically, speciation by polyploidy is seen.

If the genes are in the same chromosome, and they are located closely, there is a high chance that they were inherited together. This is called 'linkage.' Mendel accidentally selected seven characteristics in the pea plant, which were located in seven different chromosomes. It was actually fortunate for science—if not, his experiments would have confused him. During meiotic division, there is crossing over of segments of chromosome pairs. Hence the genes, which are located adjacently, show stronger linkage.

6
Lucy to Darwin

The evolution tree of humanoids (approximate):

Orrorin tugenensis	(6 million years ago)
Ardipithecus ramidus	(4.4 million years ago)
Australopithecus anamensis	(4.2 to 3.9 million years ago)
Australopithecus afarensis	(3.6 to 2.9 million years ago)
Kenyanthropus platyops	(3.5 to 3.3 million years ago)
Australopithecus africanus	(3 to 2 million years ago)
Australopithecus aethiopicus	(2.7 to 2.3 million years ago)
Australopithecus garhi	(2.5 million years ago)
Australopithecus boise	(2.3 to 1.4 million years ago)
Homo habilis	(2.3 to 1.6 million years ago)
Homo erectus	(1.8 to 0.3 million years ago)
Australopithecus robustus	(1.8 to 1.5 million years ago)
Homo heidelbergensis	(600,000 to 100,000 years ago)
Homo neanderthalensis	(250,000 to 30,000 years ago)
Homo sapiens	(100,000 years ago to present)

Charles Darwin published *Descent of Man* in 1871, twelve years after the publication of *The Origin of Species*. He stated in a letter to Lyell, "To show how minds graduate, just reflect how impossible every one has yet found it, to define difference in mind of man and lower animals: The latter seem to have very same attributes in much lower stage of perfection than lowest savage. I would give absolutely nothing for theory of natural selection, if it requires miraculous additions at any one stage of descent."

Wallace was not so convinced, and he felt that natural selection could not account for the evolution of the human brain. He felt that a big brain was not an essential commodity for catching prey.

Paleo-anthropology is a field that deals with the evolution of the humans. Conventionally, fossils were used as evidence and for understanding evolutionary trends. In recent times, DNA analysis has played a major role in giving an insight into the understanding of evolution. This method, known as the 'molecular clock', calculates the time required for accumulations of divergent mutations of lineages. By this method, we can find out the date of the diversion of two species. It is found that the closest relatives of humans are baboons. The similarity between baboons and humans in their DNA sequence is about ninety-five to ninety-nine percent—we share ninety-five to ninety-nine percent of our genes with baboons.

It has been derived from genetic studies that, during the late cretaceous period, about eighty-five million years ago, primates emerged from mammals. About fifteen to twenty million years ago, the family Hominidae diverged from the gibbons (Hylobatidae family). Around fifteen million years ago, Ponginae (orangutans) diverged from Hominidae. One of the important characteristics of the hominid is bipedalism. Bipedalism might have started with sahelanthropus or orrorin (Sahelanthropus tchadensis—seven million years ago; Orrorin tugenensis or 'millennium man', Kenya—six million years ago). Ardipithecus is supposed to be the full bipedal. During this period—about four to six million years ago—the gorilla and the chimpanzee diverged. Sahelanthropus or Orrorin is supposed to be the common ancestor of the gorilla, chimpanzee and hominid.

Fig 21   *Human evolution*

Two branches emerged from Hominoidea—Hominidae and Hylobatidae. From the former emerged Homininae and Ponginae (gibbons). From Homininae diverged Hominini and Gorillini (gorillas), and Hominini diverged to homo (humans) and Pangenus (chimpanzees and bonobos).

Genetic data can give us a great deal of information about our ancestry, migration, separation and relationships with other species. Mitochondria is outside the nucleus and is transferred only maternally. From the study of mitochondrial DNA, it has been estimated that the last female common ancestor whose genetic marker is found in all modern humans, must have lived around 200,000 years ago.

Ardipithecus

Australopithecus

Kenyanthropus platyops

Homo: Homo habilis, Homo ergaster, Homo erectus, Homo rudolfensis, Homo antecessor, Homo heidelbergensis

Homo sapiens idaltu

Homo sapiens sapiens

Ardipithecus

Ardipithecus is supposed to be the ancestor of humans ('ardi' means ground, 'ramid' means root and 'pithecus' means ape). Two species of ardipithecus have been described in scientific literature—Ardipithecus ramidus and Ardipithecus kadabba. The former lived around 4.4 million years ago *('Fossils From Ethiopia May Be Earliest Human Ancestor' by Perlman, David; July 12, 2001; National Geographic News)* and the latter about 5.6 million years ago. This belongs to an extinct genus Hominine. They lived during the late Miocene and early Pliocene period in Ethiopia. They may have been very similar to chimpanzees in their behavior.

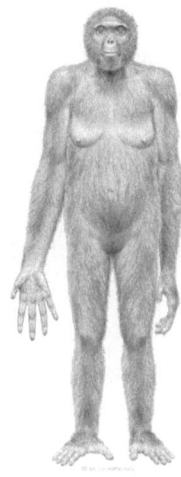 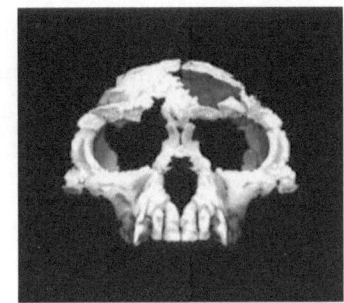

Fig 22 *Ardipithecus ramidus*

Fossil remnants of Ardipithecus ramidus were discovered in the Middle Awash river valley of Ethiopia in 1992–93, which included the skull, mandible, teeth and arm bones. More fragments were recovered in 1994, which amounted to about forty-five percent of the entire body's bones. The basic physical features were the following: Grasping big toe, which facilitated arboreal locomotion (however, they were better suited for walking in comparison with chimpanzee, but less efficient than of hominids). The canine teeth were reduced in size (upper canines were not as sharp as that of chimpanzees). Sexual dimorphism both in the body size, and the size of the canine is minimal, in comparison with apes, and it suggests that they were less aggressive. Teeth size also suggests that they were omnivores and frugivore.

Reduced sexual dimorphism indicated reduced male conflict, increased parental investment and better pair bonding. It could be considered that behavioral changes and better cooperation among the individuals had occurred even before the brain started to become larger. The brain size was only 300 and 350 cc (slightly smaller than chimps or bonobos and much smaller than Australopithecines like Lucy, which was between 400 and 550 cc). They would have been about fifty kilos by body weight. Their face was more prognathic (protruded jaw) than humans. The toe and pelvic structures suggest that they walked upright. Ardipithecus kadabba is considered the ancestor of Ardipithecus ramidus. Many researchers consider Ardipithecus as the ancestor of Australopithecus.

Australopithecus afarensis

This hominid lived between 3.9 and 2.9 million years ago. They are more closely related to the genus Homo than any other primate. A partial skeleton of a female Australopithecus afarensis, found by Donald Johanson and his colleagues in 1970, was famously named 'Lucy' in memory of the Beatles' song 'Lucy in the sky with diamonds.' The fossils of Australopithecus afarensis were predominantly found in Eastern Africa.

In comparison to the modern human, their canines and molar teeth were larger, but smaller than the modern and extinct great apes. Their brain size was 380 to 430 cc and they had a prognatic face. Though they were bipedal, they were partially arboreal, which is inferred by the structure of the scapula, feet, shoulder joints and hands. The structure of the fingers suggests their ability to grasp the branches of the trees. However, loss of abductablity of the big toe suggests their inability to grasp with the foot.

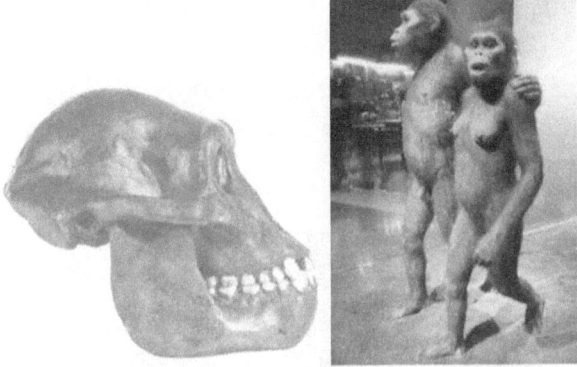

Fig 23 *Australopithecus afarensis*

Their pelvis was more like that of the human—wide, with laterally oriented iliac blades. Angulation of the femur and the calcaneus bone were features that favor bipedalism. It has to be noted that upright walking is more efficient and more energy saving than walking with bent knees and hips. The hands were longer than humans in comparison to the body size. Males were larger than females (sexual dimorphism). They might have lived in small family groups with one prominent male and many breeding females.

Kenyanthropus platyops

Kenyanthropus platyops was discovered in Lake Turkana, Kenya, in 1999, by Meave Leakey's team. ('Platus' means flat and 'opsis' means face.) This species had a flat face in comparison with the earlier species. The fossil was 3.5 to 3.2 million years old, and was a contemporary of Australopithecines. They shared some similarities with Australopithecus, but the differences outweighed the similarities and, hence, they were grouped into a different species. The similarities were seen in the brain size, nasal, suborbital and temporal regions. They had a comparatively flat face, (orthognathic—sub-nasal region), anteriorly placed zygoma (malar bones) and small molars.

Fig 24 *Kenyanthropus platyops*

It was very difficult for paleontologists and anthropologists to create an exact timeline of the evolution of humans. However, the finding of the Kenyanthropus skull caused many to think again, and the common ancestor was shifted from Australopithecus afarensis to Kenyanthropus platyops

Australopithecus africanus

Australopithecus africanus were hominids that lived around 3.03 to 2.04 million years ago. They were of slender build and were more similar to modern humans than the Australopithecus afarensis. A fossil was excavated from Taung, South Africa (1924), by Raymond

Dart and was called the 'Taung child.' The other sites of fossils were Sterkfontein (1935), Makapansgat (1948), and Gladysvale (1992).

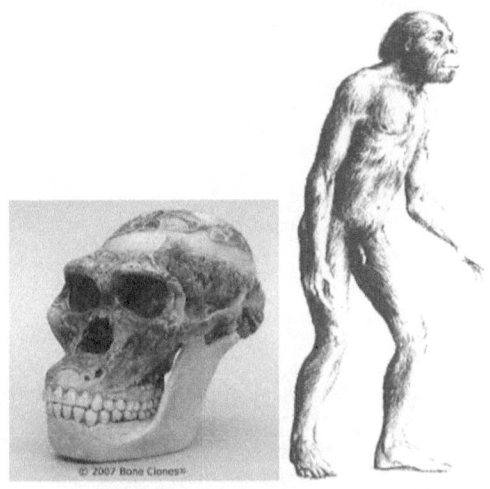

Fig 25   *Australopithecus africanus*

Their orbits, teeth and foramen magnum were very similar to humans. The position of the foramen magnum suggested an erect posture. Their arms were larger than their legs, like chimpanzees and afarensis, showing their ability to climb trees. They had a rounder cranium and the brain size was 400 to 500 cc, which is larger than Australopithecus afarensis. The pelvis was better adapted for bipedalism than Australopithecus afarensis. Sexual dimorphism in the lumbar spine was evident in the Australopithecus africanus, which is an adaptation for females.

(The wedge-shaped vertebrae in the lower back are an important evolutionary adaptation for female bipedals, for maintaining a stable posture during pregnancy. Females have three such vertebrae, while men have only two, and these together create a natural curve that helps them to bear shearing stress and increased weight in the abdomen during pregnancy. The joint's size in the lower vertebrae is larger in women than in men.)

Homo habilis

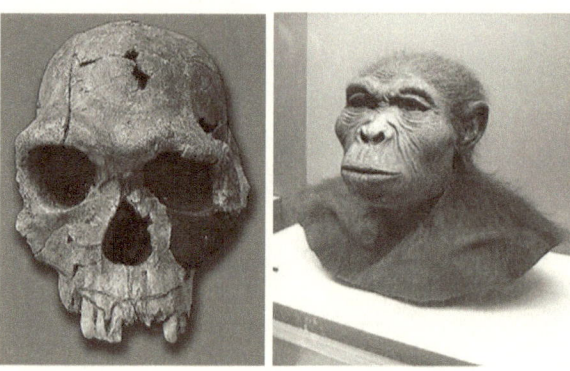

Fig 26   *Homo habilis*

Homo habilis lived about 2.33 to 1.44 million years ago, and had a brain size of about 550 to 680 cc. Homo habilis is considered as a common ancestor from which divergence took place—the modern man evolved from one of these branches. Their hands were proportionately longer than that of the modern man. The meaning of the term is 'skillful man' and they were considered to be early toolmakers, since tools were found along with their fossils. Homo habilis might have co-existed with Homo erectus, representing separate lineages or might have been its ancestor. They used the tools for scavenging rather than for defense or attack (Oldowan tools). They were on a less specialized diet in comparison with their robust relatives like Paranthropus boisei, who became extinct.

They had a larger brain than their ancestors, which resulted in remarkable changes in the shape of the skull. The size of the brain averages 610 cc (1.7 percent of the body size), the skull is more rounded and the forehead is prominent. The foramen magnum was in the center of the base of the skull indicating a straight posture. The brow ridge is smaller than that if their ancestors. Prognathism had reduced. However, many other features of the torso and the limbs were comparable to that of the Australopithecus.

Homo genus had smaller teeth than Australopithecus. The enamel of their teeth was thick, and jaws were strong. They had versatile food habits—they might have resorted to butchery of large animals, evidenced by the presence of cut animal bones with their fossils. Their jaws were smaller and more rounded (like humans) than the Australopithecus.

The features of the leg and foot indicated bipedalism. The legs were shorter and hands were longer in comparison with humans. Finger bones were a little curved (for grasping). This species was in-between humans and Australopithecines. The females were about 110 cm and males were 130 cm tall.

The grassland where they lived became cooler and drier. This climatic change might have stimulated a change of lifestyle to scavenging and use of tools. It is assumed that Homo habilis was more advanced in intelligence and culture and social organizations than the Australopithecus.

Homo rudolfensis

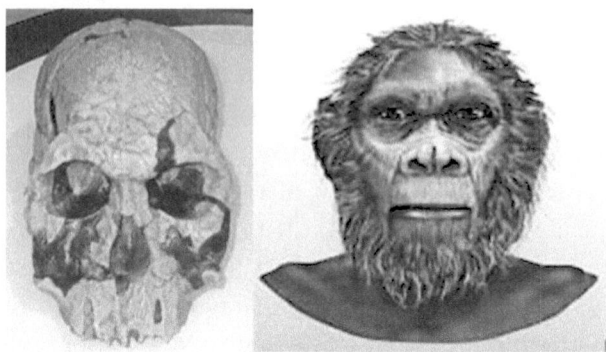

Fig 27 *Homo rudolfensis*

The extinct species of Homo genus lived around 1.9 million years ago. Richard Leaky and Meave Leaky discovered the first fossil in 1972 on the eastern side of Lake Rudolf (now known as Lake Turkana) in Kenya. The brain size is calculated to be around 700 cc. It is still a controversy that whether Homo habilis and Homo rudolfensis should be classified into one or two different groups. On comparison, the skull is rounder and the face is less prognatic than the older finds of Homo habilis. There was strong dimorphism for this species. It is considered that Homo habilis, Homo rudolfensis, Homo erectus, and Paranthropus boisei co-lived in the East African area. This finding challenges the idea that humans evolved one after another in a single lineage.

Homo ergaster

The meaning of this term is 'working man', and this species lived between 1.8 and 1.3 million years ago. The widely accepted concept is that they are the direct ancestors of Homo heidelbergensis, Homo sapiens, Homo neanderthalensis and Asian Homo erectus. They made better tools (Acheulian tools) than Homo habilis, their predecessor. Sexual dimorphism had reduced from ancestors to predecessors, from the Australopithecus onwards. However, it was more than the modern humans. The reduction in dimorphism is attributed to the diminished competition between males for mates. It is considered that they were the ones who used fire. They had a better social life and possessed linguistic abilities, though their word power was much less than that of the modern man. Their brain size averaged about 850 cc.

Australopithecus afarensis, though bipedal, was almost ape-like except for its bipedal nature with funnel-shaped rib cage, large face in proportion to the cranium and flat feet and long toes. In Homo ergaster, similarity to the modern man is much more, except for the skull and size of the face. They had short toes and arched feet, a bipedal characteristic.

This species is considered as sub-species of Homo erectus. Homo ergaster used more sophisticated stone tools than its ancestor Homo habilis. The tools were advanced from the Oldovian to Acheulean bifacial axes. Sexual dimorphism has reduced, but it was greater than modern humans. Their social organization was advanced.

Homo erectus

The meaning of this term is 'upright man.' They lived during the most of the Pleistocene period—the earliest lived about 1.9 million years ago, and the latest about 1,43,000 years ago. They originated in Africa and spread to India, China, Sri Lanka, Java and Georgia. The 'Java man' (1890s) and 'Peking man' (1920s) were classic representatives of this species. Their relation to other Homo species is not yet conclusive and is still being debated. Homo erectus' cranial capacity was greater than that of Homo habilis. The earliest ones (Damanisi specimen) had 850 cc, but the later Java species had 1,100 cc of cranial capacity. This is only about sixty percent of the human brain, but about fifty percent larger than the Homo habilis' brain. They had an orthomorphic face, less sloping forehead, less prominent cheekbones and large brow ridge. The size of the teeth was between the Australopithecus and modern man. They had a history of more than a million years and there was definitive increase in brain size and reduction in the tooth size during this period, which is evinced from the fossils. The back muscles of the head were stronger than modern humans. They had a shallow forehead and the skull was elongated, which suggests that they had narrow temporal and frontal lobes of the brain. This means that their mental ability was less than humans. Males were about twenty-five percent larger than females (sexual dimorphism). It is also suggested that they used rafts to travel. Their body proportions were very similar to that of modern humans. Their erect posture, long legs and shorter arms (compared to their torso) indicated the loss of arboreal ability and the ability to walk. They were adapted to terrestrial life. They consumed lots of meat, which helped them gain calories easily, and which increased the brain size. This also helped the women to have more children. Moreover, they lived in groups and took care of the weaker and older members, and the females during pregnancies. This increased the population and initiated the migration and spread of the species to other distant areas of the globe.

Homo erectus and ergaster must have been the first hunter-gatherer tribes, and erectus and ergaster were probably the early ones who resorted to group hunting. They cared for their weak companions and the elderly. Similarities of the cervical vertebrae with Homo sapiens and the presence of Broca's area suggest that they had a fairly well developed language. Homo erectus must be the most long-lived species, having existed for about one million years; in contrast, Homo sapiens has a history of only 2,00,000 years.

Saharan pump theory: About 1,33,000 and 1,22,000 years ago, the southern parts of the Sahara-Arabian desert had abundant rainfall for thousands of years and, later, reverted to arid conditions. This allowed migration of biota between Eurasia and Africa. These climatic changes isolated the populations, which led to adaptation changes and might have given rise to 'allopatric speciation', otherwise known as 'geographic speciation' ('allos' means other and 'patris' means paternal). This happens when the same species is separated geographically for a long period, so that adaptation changes interfere with genetic interchange, and help form a new species.

Homo erectus were efficient runners like modern humans. (They might have become more hairless—hair prevents evaporation of sweat, so hairlessness helped them to remain cool during exertion.

Other mammals keep cool by panting; since they cannot pant while running, their bodies became overheated fast. Thus, hairlessness was an advantage for humans while hunting.) It is a neotenous character as well, which is an evolutionary tendency that maintains childlike characteristics. However, they retained the hair on the head to protect the cranium from the sun as they started walking upright (or is it similar to the plumage of the peacock?). The sweat theory has got its downsides and controversies as well.

Bipedal walking released the hands from the ground for various other purposes, such as tool making. Their capacity for tree climbing suffered. There are certain adverse effects in bipedal walking: All the weight of the body has to be borne by the backbone and lower limbs. The compressive force was coped with by the double curvature of the backbone and intervertebral disks. The legs were brought under the body with the knees becoming closer and articulation of femur with pelvis also changed.

Homo floresiensis

Peter Brown, an Australian anthropologist, and his team found the fossil remains of a dwarf Homo erectus in the island of Flores, Indonesia, in 2003. The adult female was about thirty years old, was three feet and six inches tall, weighed thirty kilos and had a brain size of 380 cc. The bones of eight other individuals were also found on the island. They lived between 800,000 to 12,000 years ago. The dwarfness of this species was attributed to the 'island dwarfing' or 'insular dwarfing' phenomenon. Dwarf elephants were also present on that island.

Scarcity of food could be one of the reasons for their dwarfness, as suggested by Brown et al. Some suggest that it could be due to microcephaly, a pituitary gland disorder (Larson syndrome) or due to hypothyroidism due to lack of iodine in the diet. They are the smallest in the genus Homo. A volcanic eruption about 12,000 years ago wiped out this Homo species as well as the local fauna, including the stegodon or dwarf elephants.

(Insular dwarfing is a form of phyletic dwarfing, by which the size reduces over a number of generations when the population range is small, especially on islands. This process has taken place several times in history. Examples are dinosaurs such as europasaurus, and dwarf elephants from Crete Island. Scarcity of food is supposed to be the main cause for reduction is size, as smaller animals need less food and have a better chance of survival. Additionally, a smaller size is advantageous from the standpoint of reproduction. Island gigantism is another phenomenon seen among smaller animals that do not have any predators. The dodo is an excellent example—its predecessors were normal-sized pigeons. Another example is the large rats that co-existed with Homo floresiensis and dwarf stegodones on the island of Flores.)

Homo antecessor

Often, there is controversy and debate regarding the position of a particular fossil in the evolutionary ladder. This was one of the earliest human species in Europe, and lived around 120,000 to 80,000 years ago. Some consider it a separate species. Fossils of this species were found in Spain, England and France. The fossils found in the Atapuerca Mountains, Spain, indicates that this species might have practiced cannibalism. They were about 1.6 to 1.8 m tall and the males weighed about ninety kilos. The brain size was 1,000 to 1,150 cc. They had a flattened face, and a deep fossa between the nasal cavity and cheekbone. They had a low forehead and small lower jaw, and the chin was not well developed. Their teeth were like that of the Homo erectus. The presence of canine fossa and the shape of the nasal region were like that the modern man. Hence, this species is considered to be an evolutionary link between Homo ergaster and Homo heidelbergensis.

Some researchers are of the opinion that they had a hearing frequency similar to that of modern humans, and considered that they used symbolic language and had the capacity to reason. Another similarity shared with Homo sapiens was their pattern of development, though it might have been faster. They were excellent hunters—evidenced by the abundance of stone tools and animal bones found along with the fossils. Some researchers consider this species as the last common ancestor between the neanderthal man and modern humans.

Homo heidelbergensis

This is an extinct species, considered to be the direct ancestor of Homo sapiens, and maybe the neanderthal man and denisovans as well, who lived in Europe and Central Asia respectively. The fossil remnants of this species were discovered near Heidelberg in Germany in 1907. Afterwards, many fossils of this species were discovered in Africa, Europe and Asia. They lived around 1,300,000 to as recently as about 200,000 to 250,000 years ago. They might have descended from the Homo ergaster. They had a cranial volume ranging from 1,100 to 1,400 cc (the brain size of modern humans averages 1,350 cc). The cranium was more elongated than modern humans. The average height was about 1.75 m and average weight was sixty-two kilos. The females averaged 1.57 m in height and fifty-one kilos in weight. Some of the members were very tall (even seven feet) and their fossils were found in South Africa between 500,000 and 300,000 years ago. The tools used by Homo heidelbergensis were more advanced. Fossil findings from Atapuerca of Spain suggest that they buried their dead—probably the first species to do so. The anatomy of the ear suggests that their hearing quality was very similar to that of humans, and that they could differentiate different sounds. Hence, it is highly probable that they have a reasonably developed language capability (Homo erectus was probably the first species that started speaking).

Fig 28   *Homo heidelbergensis*

The divergence of Homo heidelbergensis started after their spread to different areas. The isolation of these groups for a long time and adaptation might be the reasons for the divergence into different species—Homo sapiens, neanderthals and denisovans in in Africa, Europe and Asia respectively (allopatric speciation). A DNA analysis of Homo neanderthalensis and sapiens indicate that they shared a common ancestor about 400,000 years ago. Homo neanderthalensis retained almost all the characteristics of Homo heidelbergensis—such as a high brow ridge, sloping forehead, lack of a prominent chin and prognathic face. The brain size was almost 1,600 cc (larger than Homo sapiens). Homo sapiens had a high forehead housing the frontal lobe of the cerebral cortex, supposed to be the seat of intelligence. The brow ridges were not prominent, and they had a definitive chin and flat or orthognathic face. The neanderthals had a parabolic jaw with smaller teeth than the previous forms but larger than the modern humans.

These species were good hunters and fossil evidence shows that they used poles with weights tied around the end portions so that they can be used as javelins. They were group hunters, and the fossil finds suggest that they hunted routinely, and might have used the hides as clothing especially in cold weather. Professor Clive Gamble of the Center of Archeology of Human Origins at the University of Southampton, England, felt that many of the hunting tools made by them were not actually used, and, hence, making them was a part of a ritual or to show off who they were. It is also thought that they buried their dead, which was a part of their complex mental development and symbolic thinking.

Homo sapiens idaltu

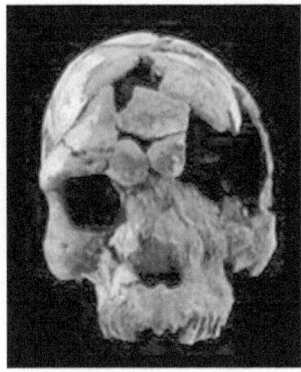

Fig 29   *Homo sapiens idaltu*

In 1997, Tim White discovered this species at Hert Bouri near the Middle Awash site of Ethiopia's Alar triangle. This is supposed to be a sub-species of the Homo sapiens that lived around 160,000 years ago. The meaning of the term 'idaltu' is 'elder' or 'first born.' Their brain capacity was 1,450 cc. The discoverer considered it an intermediate between anatomically modern humans (Homo sapiens sapiens) and the archaic human. They had a comparatively larger and rounded skull and a flat and small face in comparison with the archaic human, but they had a larger face than modern humans. The finding of this fossil was strong evidence against the theory that modern humans evolved in many places around the world. This find is also considered as evidence that modern humans first emerged in Africa. These humans pre-dated neanderthals and hence could not have descended from them. (Neanderthals branched out about 300,000 years ago and became extinct about 30,000 years ago.)

Mitochondrial DNA analysis is an excellent tool to analyze the ancestry, as it is inherited only maternally. This analysis shows that the modern human's ancestral mother lived somewhere in Africa about 150,000 years ago. The same conclusion was arrived at by the researchers who studied the 'Y' chromosome, which is of paternal inheritance. Homo idaltu, also known as the 'herto man', lived at almost the same time in Africa. So, it is postulated that the herto man might be the immediate ancestor of the modern man.

Homo neanderthalensis

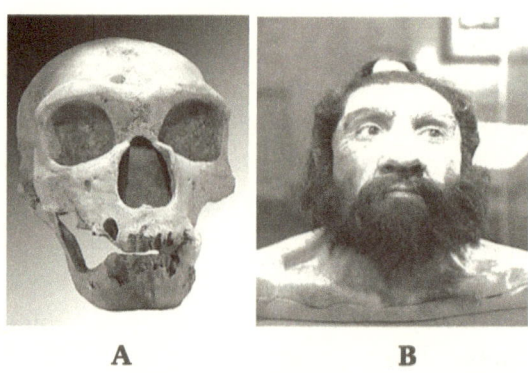

A **B**

A: *Neanderthal skull at La Chapelle-aux-saints*

B: *reconstructed head of a male neanderthal who lived 70,000 years ago*

Fig 30 *Homo neanderthalensis*

Neanderthals, an extinct species, lived in Eurasia and in the region from western Europe to Central and Northern Asia. Some scientists consider them a sub-species of Homo sapiens and, hence, also called them Homo sapiens neanderthalensis. They were discovered in the Neanderthal valley in Germany in a limestone quarry in August 1856 (three years prior to Charles Darwin's publication of *On the Origin of Species*). However, prior to this, the skulls of this species were discovered in 1829 in the Engis caves in Belgium and at Forbes quarry at Gibraltar in 1848. Later, more than 400 bones of neanderthals were discovered.

Analysis of the DNA of modern humans and neanderthals show a difference of only 0.12 percent (Eran Meshorer, Liran Carmel, et al., 2014). A comparison of the DNA suggests that the modern humans and neanderthals diverged from Homo heidelbergensis about 350,000 to 400,000 years ago. The Homo heidelbergensis spread across Africa and Eurasia, and, probably, the ones in Africa developed into modern humans and the ones in Eurasian developed into neanderthals.

Neanderthals were more robust and stronger than modern humans. They were of the same height and weight, with sexual dimorphism like that of Homo sapiens. A distinctive feature was that their cranial capacity was more than that of modern humans and averaged 1,600 cc. Neanderthals were mainly carnivores, but Henry A.G. et al (in 2010) found microfossils of cooked vegetable matter in the calculus of neanderthal teeth. This suggests that they were not only omnivores, but also used fire to cook. They made advanced tools and built houses using animal bones. They also made boats and were ancient mariners. A find in the Zagros Mountains in Kurdistan showed that one of the skeletons was buried with flowers, which suggests that they had burial rituals. They probably had a spoken language.

Though the neanderthals and ancestors of modern humans lived in the same locality, from the findings of the analysis of DNA, it is postulated that there might not have been inter-breeding between these two species. Genome analysis points out about thirty differences between modern humans and neanderthals. One is RPTN—a gene that encodes reptin, which is an extracellular epidermal matrix protein. Neanderthals were missing this protein, which helped them to adapt better to cold climate but made them less resistant to diseases.

It seems that the neanderthals existed till about 30,000 years ago, which means that they co-existed with the modern humans. The neanderthals' body was better suited for cold climate. Weather fluctuations, which occurred about 55,000 years ago, might have been one of the causes for their extinction. Another reason suggested was that modern humans moved into their area, and a violent conflict ensued and the neanderthals were displaced. Neanderthals needed more energy than modern humans for sustenance due to their large brain size and anatomical structure (they needed more energy than any other Hominid species). The lean periods might have affected them adversely. Another theory is that, due to inter-breeding, they might have been completely absorbed into the cro-magnon population.

Cro-magnon Man

The cro-magnon were the early European modern men, the Homo sapiens sapiens dated 43,000 years ago. They were robust and powerful. Their brow ridge was not very prominent, they had a prominent chin and a straight forehead, with a brain volume of 1,600 cc (larger than the average humans). A complete skeleton that was around 33,000 years old was discovered in 1823 in a cave burial in Grower, South Wales, the United Kingdom. The burial was a ceremonial one, as there were personal decorations and numerous tools, suggesting religious practices.

Fig 31 *Cro-magnon man*

(Modern humans have evolved about 100,000 to 200,000 years ago in Africa. Then, they spread to Eurasia about 60,000 years ago. Migration to Europe started about 45,000 years ago).

Cro-magnons were tall, with an average height of 176.2 cm. Some of them were as tall as 195 cm. They had an orthognathic wide face, high forehead, and large nose. Their vocal apparatus was similar to that of modern humans, which means that they could talk. The flint tools found with the remains, their paintings, pierced bones and shells for body ornaments suggest an advanced 'Aurignacian' culture compared to the previous forms.

(Aurignacian culture belongs to the Upper Paleolithic period, which was located in Europe and Southwest Asia. The name originated from Aurignac in France. This culture lasted from 47,000 to 41,000 years ago.)

Evolution to Homo sapiens

Australopithecus afarensis (lived 3.5 to 3 million years ago)

Australopithecus africanus (lived 3 to 2.4 million years ago)

Homo genus (Oldowan tools)

Homo rudolfensis   Homo habilis

(Lived 2.2 to 1.8 million years ago)   (Lived 2.2 to 1.6 million years ago)

Homo ergaster (lived 1.8 to 1.2 million years ago)

[Acheulean tool]

Homo erectus (lived 1.3 to 0.2 million years ago [Fire]

Homo antecessor (Homo heidelbergensis) (Lived 1 to 0.2 million years ago)

Homo sapiens (Modern man)   Homo neanderthalensis

(0.2 to 0.0 million years ago)   [Art, built shelters]

(Lived 0.4 to 1 million years to 35,000 years ago)

Climatic cooling during the late Miocene period (6.0 to 5.3 million years ago) might have triggered speciation of the Hominin super family. This period is marked by speciation of other mammalian families as well. Diversification of humans and chimpanzees took place sometime during the late Miocene period. This period was marked by climatic changes compounded by global cooling and drying. However, lack of sufficient fossil records during this period leaves questions over the veracity of this generally accepted postulate.

These geological events caused the common ancestors of humans and chimpanzees to split into rainforest dwellers of West Africa and open, dry habitat dwellers of East and (perhaps) North Africa. The former evolved into arboreal modern chimpanzees and the latter into modern humans. The period of separation of the humans and the chimpanzees from their common ancestors is debatable. Various molecular studies and gene analysis vary from four to 10.5 million years ago.

(In 1655, Isaac de la Payrere from France discovered stone tools used by primitive men. He claimed that these tools belonged to men who lived before biblical Adam. His findings were condemned and the church authorities burned his books. By the late 18th century, paleo-anthropology became a scientific discipline.)

As the ancestors of humans climbed down from trees and started walking on their hind limbs, their forelimbs became free and took over many of the functions of the jaws, such as fighting, carrying, holding and tool making. The hominid lineage probably started when the pre-humans started throwing stones and swinging clubs against their adversaries. Prowess in throwing and clubbing was a definitive advantage in natural selection (power grip and precision grip for throwing and clubbing respectively). Primates had a diminutive thumb and elongated curved fingers, which were essential for their arboreal way of life. In humans, the fingers got shortened and straightened and became larger and muscular. Humans developed a fully opposable thumb. When the human fingers are flexed, the rotation is on a central axis so that the fingertips can meet the tip of the thumb.

Humans and chimpanzees have more similarities than differences—they have the same number of bones and muscles and the basic framework is very similar, except for the proportions. These proportions are adaptations for varied life styles. Chimpanzees have long powerful arms, and their toes are long with opposable big toes adapted for climbing. In humans, the toes are arranged straight and arms are shorter than legs, which are meant for straight walking. The thumb is opposable to enable tool making. The chimpanzee's rib cage is funnel-shaped, and barrel-shaped in humans. This represents different eating habits. The rib cage generally protects the lungs, heart, liver and other organs of the thorax. The chimpanzee's rib cage is wider below to accommodate the large belly, which is characteristic of plant eaters. Humans have the small belly of a flesh eater. The lumbar spine is straight in chimpanzees and curved in humans. The pelvis is long and narrow in chimpanzees, while humans have a wide pelvis. The chimpanzee's knees are wide apart, while humans' knees are closer together. The cranium is smaller in comparison to the face for chimpanzees and larger in humans. Dimorphism is predominant in chimpanzees as well as gorillas, and this gradually reduced as the hominids ascended the ladder of evolution.

Though modern humans and apes show stark differences in their posture, structure and morphology, the differences were less with the earliest Hominins. The differences were mainly due to bipedalism (walking upright on two legs), which brought the foramen magnum forward. The pelvis became bowl-shaped as it shortened and broadened. Many factors were responsible for facial changes—receding jaws and expanding skulls led to anthropologists believing that the 'brain case expanded at the expense of the jaws.' Several environmental, developmental and evolutionary factors played a role in the changes that took place in the physical characteristics of humans.

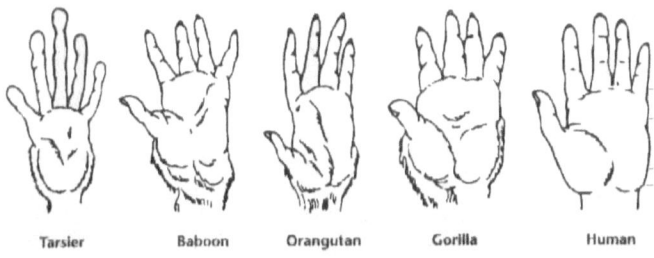

Tarsier   Baboon   Orangutan   Gorilla   Human

Fig 32 *The opposable thumb gave the hand more dexterity and helped in making and using tools*

The modern man is scientifically known as 'Homo sapiens sapiens.' Living humans, all human ancestors and many extinct members of the Australopithecus belong to the 'Hominidae' family. Our immediate biological cousins are Cercopithecoidea (old world monkeys and apes) and Pongidae (chimpanzees, gorillas and orangutans). It is generally accepted that we are not the direct descendants of any of the existing monkeys or apes, or even the old world monkeys. We had a common ancestor.

In 1856, a strange skull was discovered in the Neanderthal valley in Germany. The skull belonged to a hominid later named as Homo neanderthelensis.

Ernst Heinrich, a German Scientist, opined that the skull was half-human and half-ape. (Though Charles Darwin had completed his voyage on the Beagle, his book, *Origin of Species* was not published at that time; it was published on 4 November, 1859.)

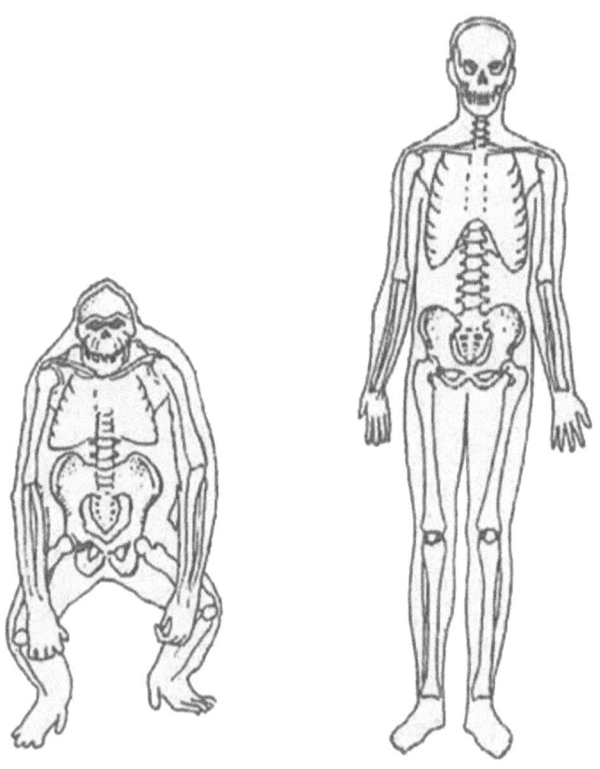

Fig 33 *The straight posture of the humans made the face straight and pelvis short and bowl-shaped; the foramen magnum was shifted forward*

In 1890, a Dutch physician named Eugene Dubois opined that the skull found at the banks of a river in Java was the link between man and ape. Johansson is credited with the finding of the fossil skeleton of 'Lucy' (Australopithecus afarensis) in Ethiopia—the upright walking mother of all modern humans. Members of the family Hominidae are bipedal primates.

From Australopithecus onwards, the evolution of humans took place at a breakneck speed. The reasons attributed to this are many: Early humans were a restless species—they had a tendency to migrate and were subjected to varied climatic, geographical and environmental conditions, which necessitated adaptation. This acted as an impetus for evolution through natural selection. During the last two million years, there was a co-existence of different species of the hominid species.

7
Beauty, a Lure

Asexual reproduction is a type of reproduction by which there is no fusion of gametes. The offspring will have the exact copy of the genes of the parent. It is prevalent in single-celled organisms such as archaebacteria, eubacteria, and protists. Many plants and fungi reproduce in the same way. Asexual reproduction provides fast regeneration and provides rapid propagation. It is rare in animals and in many multicellular organisms. Asexual reproduction offers short-term benefits, while sexual reproduction aims for genetic diversity, which is inevitable for better adaptation to the environment and evolution.[23] If the gene is too selfish, there is no reason for development of sexual reproduction. In sexual reproduction, reduction division (meiosis) takes place, resulting in recombination and repair of the genes

The simplest form of asexual reproduction is binary fission, where the parent cell divides into two daughter cells. Budding, vegetative reproduction, spore formation, fragmentation and so on are certain other types. Parthenogenesis is a type of asexual reproduction where an unfertilized egg develops into a new individual. This is observed in many plants, invertebrates (such as water fleas, rotifers, aphids, stick insects, some ants, bees and parasitic wasps), and vertebrates (such as some reptiles, amphibians, and rarely birds). 'Facultative parthenogenesis' is the term for the occurrence when a species that normally reproduces sexually undergoes asexual reproduction. Facultative parthenogenesis occurs in certain sharks and reptiles as a response to the absence of males in the habitat. This is in contrast to 'obligate parthenogenesis', where the organism reproduces exclusively by asexual means. There are around eighty species of reptiles, predominantly lizards, which are unisex, where the males have no role in reproduction. Several experiments are being conducted on parthenogenesis. (*Parthenogenesis in humans: On 26 June, 2007, the International Stem Cell Corporation (ISCC), a California-based stem cell research company, announced that their lead scientist, Dr. Elena Revazova, and her research team were the first to intentionally create human stem cells from unfertilized human eggs using parthenogenesis.*)

A South Korean scientist named Hwang Woo-Suk claimed that he and his team had extracted stem cells from cloned human embryos—a result that was fabricated. On 2 August, 2007, after many independent investigations, it was revealed that the discredited Hwang Woo-Suk had unknowingly produced the first human embryos resulting from parthenogenesis. In such a situation the role of males in reproduction becomes insignificant.

Bacterial conjugation is an equivalent of sexual reproduction, as there is transfer of genetic material (plasmid or transposon) to another cell. This mechanism was discovered in 1946 by Joshua Lederberg and Edward Tatum (*Lederberg J., Tatum E.L.; 1946*). This is one of the methods by which bacteria become resistant to antibiotics and achieves the ability to use new metabolites.

Transformation is the absorption of genetic material from the surrounding environment without actual contact with another cell. Transduction is the injection of foreign DNA by a bacteriophage virus into the host bacteria. Molecular biologists use transduction technique to transfer foreign gene to the host genome.

Evolution of the reproductive system started from binary fission through transformation, transduction and conjugation to sexual reproduction.

Sexual reproduction has certain definitive survival advantages. About 99.9 percent of the eukaryotes reproduce sexually. Sexual reproduction started around 1.2 million years ago. The pleasure principle as a reason for the development of sexual reproduction is flouted as the neurons for experiencing the pleasure appeared long after the development of sexual reproduction. Sexual reproduction is a costly affair for the organism, as it has to undergo meiosis and find a partner to share the gamete. Another important aspect is that only half the genes are transferred by one partner. However, there are certain important survival benefits. The most important one is that genetic variability is introduced, which has adaptive advantages. Some of the genes, which are suppressed, could express themselves after generations. Recombination of genes can counter deleterious mutations and helps to repair damages within the genes. Sexual reproduction stimulates the production of more complex organisms, thus improving the survival strategy. In 1889, August Weismann proposed that creation of variation among siblings is the main advantage of sexual reproduction.

Charles Darwin wrote in *The Origin of Species*: "It is interesting to contemplate an entangled bank, clothed with many plants of many kinds, with birds singing on the bushes, with various insects flitting about, and with worms crawling through the damp earth, and to reflect that these elaborately constructed forms, so different from each other, and dependent on each other in so complex a manner, have all been produced by laws acting around us."

According to Michael Ghiselin (*The Economy of Nature and the Evolution of Sex*, 1974), diverse siblings with varying genome would be able to get more food as they use different methods to obtain it. It has been noticed that a majority of animals that produce small siblings reproduce asexually while the ones that produce larger ones reproduce sexually. Gorelick and Heng are of the opinion that sex acts as a sieve, preventing major variations like chromosomal rearrangements but allowing minor changes at the gene level.

And Beauty

"If everything were cast in the same mold, there would be no such thing as beauty," said Charles Darwin. Sexual selection is a natural phenomenon among the individuals of the same species for the selection of a mate. Darwin considered this an important factor in evolution. Likewise, in the struggle for existence, there is a struggle for a partner, and the successful one gets more mates and offspring than the unsuccessful ones.

In social life, individuals fight for mating rights and supremacy. There is a pecking order among poultry.

Beauty, a Lure

Social animals, including humans, live in groups. It is an effective survival strategy, which is symbiotic. Group living has other advantages such as predator protection, better ability for forage for food, and sharing of information. At the same time, predator animals can hunt large animals as a group. Large groups, however, have disadvantages such as the spread of communicable diseases and competition for food and mates.

There are two types of sexual struggle. The one often seen among males is the establishment of supremacy over the rival by means of physical power. In humans, physical power as well as sociological factors play a role. Next is the struggle is to charm the opposite sex. Theses struggles depend on the ratio between the two sexes.

Many animals develop certain features that are not at all important for their survival but for ornamentation, which makes them more desirable in the eyes of the opposite sex and helps them in reproductive success. Making themselves attractive to the opposite sex is known as 'intersexual selection' and defeating or intimidating the same sex is called 'intrasexual selection.'

Intrasexual selection often helps the development of many positive characteristics for evolution, while intersexual selection can go to the extremes, as many of these characteristics are more of ornamentation, which can be a burden for the species. For example, the extinction of the Irish elk (Megaloceros giganteus) was due to its huge antlers. Characteristics developed by male combatants are secondary sexual characteristics such as weapons and ornaments. Females often prefer to mate with males that have elaborate 'ornaments', such as a large body, large sex organs and other secondary sexual characteristics. Many of these characteristics show off the animals' health, fighting ability and ability to survive. Sexual selection is an important factor for diversification and evolution. Secondary sexual characteristics are often attractive and these represent health and vitality. This has given rise to the 'good gene hypothesis.' Hence survival of the fittest becomes akin to survival of the prettiest.

Peacocks have a colorful, ornate plumage, which is considered as secondary sexual characteristic. The display of the same helps the peahen to assess the quality of the peacock's genes. Birds exhibit very interesting sexual selection behaviors—color of their plumage, songs they sing, ability in nest construction, and dancing in front of the partner are certain methods used to seduce the female partner. Male birds are generally charmers. They display their beauty to charm the females. Darwin wrote: "Ornamented by all sorts of combs, wattles, protuberances, horns, air distended sacs, topknots, backed shafts, plumes and lengthened feathers gracefully springing from all parts of the body, the beak and the naked skin about the head, and the feathers are often gorgeously colored."

Males build nests, and the females inspect them and select the male that has made the most attractive nest. A striking example is that of the bowerbirds. David Rothenberg in his book, *Survival of the Beautiful*, remarked: "The male bowerbird evolved into an artist for the simple function of attracting a female to take a look at his beautiful creation." It is not only the appreciation but creation of beauty as well that is an evolutionary saga.

It has to be postulated that the appreciation of beauty might have started millions of years ago before humans evolved. Anders Moller, a zoologist, conducted a study on swallows to find out the effect of tail length and symmetry in sexual selection. He snipped off the tails of males and altered its symmetry in some, and pasted feathers in some to increase the length. He observed that male swallows that had longer and symmetric tails were preferred.

Infants are sensitive to beauty from a very young age. Psychologist Judith Langlosis is of the opinion, "We are born with preferences and even a baby knows when she sees it." They are more attracted to symmetrical patterns more than the asymmetrical ones and prefer soft surfaces to rough ones.

Fight for dominance among the males exemplifies intrasexual selection—very common among mammals. One example is the elephant seal that maintains a harem of thirty to hundred females. Males are very large when compared to the females (dimorphism). Males reach the colonies earlier than the females and fight for the harems. Intrasexual selection is prevalent in most of the animals that live in social groups. Sexual dimorphism is a difference in characteristics between males and females. This can be the difference in size, color patterns and other ornate features such as antlers in deer, tusks in Asian elephants, plumage in peacock, and ornamental plumage of the bird of paradise. Even in humans, males are often larger than females. It has to be noted that from our hominid ancestor, Australopithecus to modern Homo sapiens, dimorphism related to size is decreasing through the intermediate Homo species.

This is applicable for humans as well. Understandably, the purpose of sex is reproduction, and sex is a pleasurable experience. Though it is thought that only humans use sex for pleasure, it is now understood that many primates also engage in sex for pleasure. Instinctual component is also involved, as it is a biological need. If sex were not pleasurable, there would not have been propagation of the species.

Both chimpanzees and gorillas become fertile at intervals, and there is strife between the males for dominance and sex. The males can identify their offspring. Though they do not help in upbringing the young ones, they help them if there is a combat.

Males are stronger and larger than females in almost all mammals, and dimorphism is more pronounced in lower forms. In all mammals including humans, the society is male dominated—least of all in humans. Chastity is linked to morality, and is imposed on women by the society. In primitive societies, the strongest male (the leader) kept a harem. (Polygamy is kingly and polyandry is prostitution—undoubtedly male chauvinism.) In humans, the leader need not be physically strong or sexually attractive. Leadership depends on multitude of factors. This, along with other factors of cultural evolution, paved the way for male-female coupling in the society—marriage. The system of marriage has reduced strife between males for a mate in the society. The urge for variety (based on curiosity—an innate nature inherited from the apes, in man) pushes both men and women to adopt sneaky ways. Survival of the 'fittest' is the Darwinian norm; attractiveness too plays an important role in evolution. Good looks are considered an external manifestation of fitness and brightness. Reproduction is the only way to propagate the gene. For women (whoever seeds), half of it is her genes. For men, there is no way to know if his partner has conceived his child (DNA scrutiny being unnatural, it doesn't appeal to emotions). The hapless male can thus ensure the propagation of his gene by ensuring the chastity of his mate. It is very interesting to note that the mother and her relatives are often in a hurry to proclaim that the newborn looks like the father or his clan. This may be an unconscious/conscious effort on the part of the mother's side to erase any doubt about the fatherhood of the child.

In *The Selfish Gene*, Richard Dawkins postulated that propagation of the gene is the primary purpose of all living organisms. It amounts to saying that the living organism is only a vehicle of the gene for its propagation.

Charles Darwin considered that the male beard and relative hairlessness of humans are due to sexual selection. He also attributed the difference between races to sexual selection. Geoffrey Miller, an American evolutionary psychologist hypothesized that many of the human behavioral traits such as humor, music, visual art, verbal creativity, altruism and chivalry are the result of sexual selection. Many of these characteristics cannot be attributed to fitness, and could not be considered as natural selection. Clothing might have developed to attract the opposite sex. Ferdinand Fellman, a German anthropologist, is of the opinion that 'self-consciousness' is due to extended sexual selection, termed emotional selection, bridging the gap between animal behavior and human erotic love.

For most animals, sex is a reproductive activity called 'copulation.' Usually, for all mammals except humans, mating occurs at the time of estrus, which is the fertile period for females. However, some studies have pointed out that many primates use sex for pleasure, as they engage in masturbation, polygamy and promiscuity and may even show homosexual behavior. Among animals, the females often select the most attractive and strong male. The male who wins the fight often gets more females and can propagate his genes more.

Monogamy is rare in animals—it varies with different branches of the animal kingdom. More than ninety percent of the avian species is monogamous and they live in pairs; in primates, it is about fifteen percent and only three percent for other mammals.

The human brain consumes about one-fourth of the total energy and oxygen consumed by the human body—which is very large, considering its cost-benefit ratio for fitness. Michael A. Schillaci, in his research article, 'Sexual Selection and the Evolution of Brain size in Primates', argued: "Brain sizes among primate species are associated with monogamous mating systems, suggesting primate monogamy may require greater social acuity and abilities of deception."

It has been pointed out by many researchers that the 'ornaments' (secondary sexual characteristics) are also used for display of warning (aposematism), which can often prevent costly physical confrontation. Many opine that many of these features Darwin attributed to sexual selection are mostly natural selection.

Beauty is a universal experience for almost all animals, especially the higher ones, and is evident among birds and mammals. Beauty fosters pleasure and promotes selection in a world of sexual reproduction. Sensitivity is hard-wired into our brains and operates at a subconscious level. From childhood, we see that beautiful people get better attention and treatment, and this, to a great extent, molds our behavior.

Alfred Russell Wallace was opposed to the concept of sexual selection as he felt that the animals are far too cognitively underdeveloped for aesthetic appreciation. However, it has been well established that female birds of some species inspect the nest constructed by the males before they accept them as a sexual partner. It has also been observed that the bowerbirds decorate their nests elaborately to attract the females.

Beauty is a universal phenomenon that nature also follows with the 'Phi' proportion or 1:1.618. This proportion known as the 'golden proportion' or 'divine proportion' (Da Vinci) is abundantly observed in the arrangement of the petals of flowers, leaves, shells and so on. Many great painters, sculptors and architects use this proportion in their work. As nature itself follows this principle, which is also functionally superior, we have to postulate that lower forms as well, by default, show an affinity towards beauty and perfection.

Freud believed that beauty is derived from sexual excitement. Though the sight of genitals may excite, they are hardly beautiful. Excessive indulgences in beauty represent pathological narcissism. Narcissism is a cover for a feeling of physical inferiority.

Fig 34 *Michelangelo used the golden proportion in his painting, 'Seduction and expulsion of Adam and Eve', in the Sistine Chapel*

Fig 35 *Venus de Milo at the Louvre has been described as a "classical vision of beauty"*

'Venus de Milo' at the Louvre is considered as a "classical vision of beauty." However, one expert claimed that her "almost matronly" representation was meant to convey an "impressive appearance" rather than "ideal female beauty." Ishtar, a Mesopotamian goddess, has been associated with sexuality, love, and fertility. Physical attractiveness is desirable. However, desirability needn't always be aesthetically superior. Physical attraction is based on many social and cultural factors, such as honesty, intelligence, and social status. Subjective personal preferences also play an important role in physical attraction. Many evolutionary psychologists like Kanazawa (2011) are of the opinion that intelligence and physical attractiveness are indicators of genetic fitness.

Youthfulness, full breasts, oval and symmetrical faces, full lips and a low waist-hip ratio are certain features that make women attractive to men. Men often prefer women who are shorter than them. This may be attributed to men's inherent dominant attitude. At the same time women are attracted to men who are taller than them. Facial symmetry, broad shoulders, V-shaped torsos and masculine facial features attract women. Over the last century, men's height has increased—attributed to better nutrition. Women often prefer better-looking men who are also financially prosperous. However, beauty need not bring extra happiness.

About 70,000 Americans undergo aesthetic surgery every year. More money is spent on beauty than on education. Before the Federal Drugs Authority (FDA) controlled silicone gel, about 400 women were getting silicone implants in their breasts every day in the United States. Psychoanalyst John Gedo and Peter Kramer considered cosmetic surgery as 'cosmetic psycho-pharmacology' as cosmetic surgery changes the personality of a person. It is another method of attaining confidence and gaining better acceptance in the society.

Generally, people expect good-looking people to be good at everything, including in bed. Various people have conducted plenty of studies on the aspect of beauty and attitude. Almost all of them conclude that attractive people are always in an advantageous position, whether in getting help from a stranger, winning an argument, getting a partner, having sex, getting away scot-free from trouble and so on. Men and women use camouflaging in their dress to look more attractive. Females use padding on their bottom as well as on the bosom to look more attractive.

Large mammae of women give the impression that they can produce more milk for the offspring. For females, the breasts are large even if it is not lactating. For all the other mammals, the mammary glands are not a sex symbol and they become large only when they are lactating. Only in human females do the breasts becomes rounded at puberty. For humans, it is an important secondary attractive sex organ, though it is deceptive. In female chimpanzees and other apes, the genitals become red and enlarged when they are in estrus—this is an invitation to the males. During bipedal walking, the female genitals are mostly covered. This might be the reason for the development of secondary sexual characteristics such as enlarged breast and buttocks. Bipedalism might also be the reason for sexual acts to be conducted from the front for humans, while the rest of the mammals conduct the sexual act from behind (large buttocks could be a hindrance for proper penetration from behind). Large breasts also act as a soft padding during Intercourse and embrace from the front. The female nipple is a conglomeration of nerve endings and is hence a highly sensual area.

The ideal proportion of the breast is forty-five percent above the horizontal line through the nipple and fifty-five percent below, with the bottom part being perfectly convex. The nipple should be pointed upwards at an angle of twenty degrees. Plastic surgeon Patrick Malluci called it the 'beckoning breast.' The breasts wobble while the female walks—an invitation to the male, and a way of saying 'I have reached the age to propagate.'

Fig 36   *Piero di Cosimo, Portrait of Simonetta*

Beautiful breasts are known as 'beckoning breasts'

Large buttocks and large breasts store fat. A normal fit man has about fifteen percent fat while a female has about twenty-three percent fat by weight in their bodies.

The large buttocks and wide hips (pelvis) of a woman give the impression that she can hold a larger child with a large head in her womb. Hence, these features are considered to be attractive. Women use high-heeled footwear to give prominence for both their bosom and bottom. Men use thick-soled footwear to look taller. The waist-hip ratio is another important criterion for beauty. An hour-glass appearance is the best. Studies have shown that most of the voluptuous models have a waist to hip ratio of around 0.7. The ideal as per the phi-ratio would have been 0.618. Maybe, humans are evolving gradually to this ratio. (In the analysis of facial beauty and evolution of face, I have shown that proportion between the cranium and face reaches the golden proportion only after one million years.)

Teachers expect good-looking students to do better in their studies. Randy Thornhill and Steven Gangestad, in their study, observed that attractive men bring their women to simultaneous orgasm more often. Beauty is a great advantage in all the walks of life and, hence, it is a positive force in natural selection.

Donald Symons, an anthropologist, one of the founders of evolutionary psychology, and the author of *The Evolution of human sexuality* is of the opinion that, in all societies, females conceive sex as a service or a favor. Men are much more desirous of sex than women. Women's sexual attractiveness is dependent on physical attractiveness, but men's sexual attractiveness lies in political and economic prowess. Symons claims that this difference in the sexual psyche of males and females stems from the 'deep seated biological roles' of the humans' hunter-gatherer days. There are certain stark differences between males and females in the morphology of the face and bone. In humans, males decide (mainly) on whom to mate with. This is a notable difference from other mammals. This is the main reason for females to appear more attractive than males in humans. Females retain more of their childhood characteristics. In human males, marked changes appear in the features from puberty when testosterone, the male hormone, is pumped in to the system. Their brow ridges are larger than females, which make the eyes appear deep-set. The nose is wider and longer, the lips are thinner and the jaws are larger and square-shaped. The female face has more fat. Female eyelashes are longer and their nose is smaller with a slightly depressed nasal bridge. Males have higher metabolic rate and they have more hemoglobin in their blood.

Symons also proposed that the idea of facial beauty is averageness. He called this a brain mechanism of 'face averaging device.' However, striking beauties are never average. Edgar Allan Poe said, "There is some strangeness in the features of very attractive people, which doesn't fit into the average. There is no exquisite beauty…without some strangeness in the proportion."

It has to be stressed that childlike features such as large eyes, large cranium in comparison with the face, oval face, smooth and soft skin are attractive in an adult face. Plastic surgery to enlarge the lips and hips and to raise the eyebrows and reduce the size of the nose makes women more attractive.

"A person's face can say a lot: Helen's face is said to have launched thousand ships, while Medusa's could turn men to stone. And even today we talk about individuals with 'a face that can stop a clock.'"*(Edward Willett, October 29, 2008)*. Studies have shown that babies prefer pretty faces, and they look longer at such faces. Body odor is another important factor that attracts men and women. It is interesting to note that, in a double-blind study, the body odor of physically attractive men was more appreciated by women, though they were not informed whose scent it was. A story goes thus, that Napoleon wrote to Josephine from a camp: "I will return to Paris tomorrow evening. Don't wash."

(For portraying the world's most beautiful woman—Helen of Troy—Zeuxis selected five models and distilled features from them and synthesized the portrait).

The proportions of the body are designed by nature in such a way that it serves the requirements to exist in harmony with the surroundings. Plato opined that, "Beauty resides in proper measure and proper size of parts that fit harmoniously into a seamless whole." Throughout the Renaissance in Europe, the general concept was that carnal beauty represents inner beauty. Baldassare Catiglione wrote: "Beauty is a sacred thing… only rarely does evil dwell in a beautiful body, and so outward beauty is a true sign of inner goodness." For St. Augustine, beauty was synonymous with geometric form and balance. He thought that equilateral triangles and squares were beautiful because their parts are more even. He considered circles more beautiful, and the most beautiful is the point (.), which is indivisible and pure. Order, symmetry and definiteness were the hallmark of beauty for Aristotle. For Cicero, Plutonius and many other artists and philosophers, property of beauty was in proportions, harmony, symmetry and balance. So, the fundamentals of beauty in all things in nature are the same.

Albrecht Durer, in his four books on human proportions, considered the physical perfection of Apollo, Adam (before the fall) and Christ. He considered their perfect beauty as a sign of their divinity, while the imperfections were the sign of fall from grace. However, he was polite, and stated that, "What beauty is, I know not, though it adheres to many things." He also said, "I hold that perfection of form and beauty is contained in the sum of all men."

(One of the earliest records on proportions of the human body was one by architect Vitruvius. Leonardo Da Vinci drew the 'Vitruvian Man' in around 1490 based on the notes by Vitruvius.)

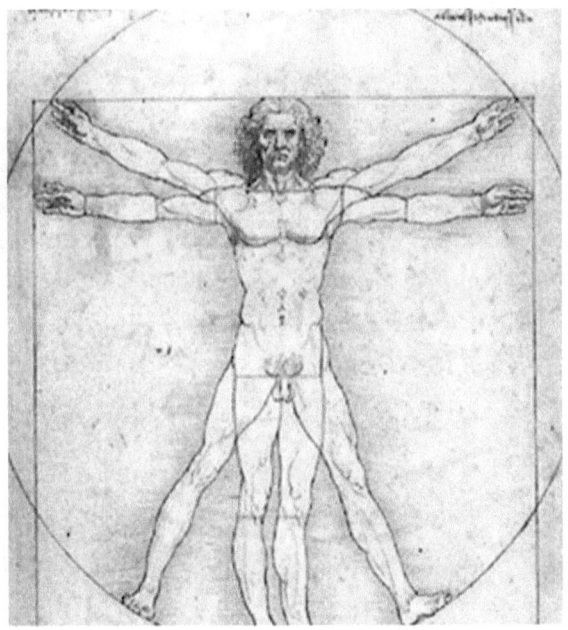

Fig 37   *Vitruvian man*

This drawing is of a man in two superimposed positions with his arms and legs apart. It is inscribed in a circle and a square. Vitruvius considered the human figure as the principal source of proportion for classical architecture. Da Vinci envisaged this anatomical drawing as a cosmography of microcosm.

Da Vinci has written: "Vetruvio, architect, puts in his work on architecture that the measurements of man are in nature distributed in this manner, that is:

A palm is four fingers

A foot is four palms

A cubit is six palms

Four cubits make a man

A pace is four cubits

A man is twenty-four palms

And these measurements are in his buildings."

"Leonardo envisaged the great picture chart of the human body he had produced through his anatomical drawings and Vitruvian Man as a cosmography of the microcosm (*cosmografia del minor mondo*). He believed the workings of the human body to be an analogy for the workings of the universe." And, "If you open your legs enough that your head is lowered by one-fourteenth of your height and raise your hands enough that your extended fingers touch the line of the top of your head, know that the center of the extended limbs will be the navel, and the space between the legs will be an equilateral triangle."

8

Homo Sapiens Divine

Beauty of the Human Face

Dutch artist and anatomist Petrus Camper devised a technique to measure facial angles from profiles. He drew a line from the ear to the lip, and another line from the most protruding point of the forehead to the most protruding point of the face (the upper lip)—the bisecting angle is the 'facial angle.' He considered Apollo and Venus (statues) as the most beautiful. He observed that these statues have facial angles of 100 degrees. They had a relatively straight profile. Most human profiles range from seventy to ninety degrees. Since the lower forms of animals have the facial angles that are more acute, he considered obtuse angles as representing beauty. He wrote: "What constitutes a beautiful face? I answer a disposition of traits such that the facial line makes an angle of 100 degrees with the horizontal." He measured the skulls of different races, and found the facial angle increased from the orangutans and monkeys to African blacks to Orientals and to the Europeans, but had not yet reached the Greek statues. He considered Europeans as being closest to the beauty ideal.

Humans' height varies from region to region. Genetics, too, play a role, but racial difference cannot be proved conclusively. It has been observed that height varies from generation to generation. This may be attributed to food habits and not to evolution. The American Indians were the tallest people on the planet. Their height averaged five feet and ten inches. This could be because they ate a high protein diet. The Europeans at those times averaged five feet and six inches. Within a short period of two centuries, their height increased by two to three inches (Hannah Holmes, *The Well Dressed Ape*).

The color of the skin was historically considered an evolutionary trait. This has been unfortunately used as a ploy to consider the dark-skinned people of Africa as an inferior race and make them slaves of white-skinned European explorers. Even now, remnants of that atrocious thought is present in certain parts of the world. It is poignant that, even in developed nations, racial prejudice raises its gruesome head—and the frequency seems to be on the rise. Skin color, in reality, is an adaptation to the environment and not an evolutionary feature. Humans evolved in the Savanna grasslands, where it is bright with sunlight. The black color of the skin was due to the melanin pigments, which protect the skin from skin cancer. Pale skin evolved in the inhabitants of the northern region, as sunlight was relatively less. Moreover, sunlight is important for vitamin D synthesis. So, pale skin absorbs more sun than dark skin. Females need more vitamin D to nourish their children. This could be the reason for their being lighter-skinned than males. The same reason can be attributed to the lack of hair on their face. The color of their hair also follows the pattern.

The face—the seat of special sensory perception—also mirrors the fine human expression. The maxims 'lose face' and 'face is the index of the mind' indicate the importance of face. During the evolutionary process, though humans lost many of the functions of the jaws, they gained deft control of the facial musculature and the smile that is unique to humans. To identify a person and to assess his/her beauty, we depend on facial features. The underlying skeletal architecture is the major decisive factor for the morphology of the face. A proper balance between the different parts of the face—forehead, eyebrows, eyes, nose, ears, cheeks, lips, mouth and chin—and the morphology and symmetry of the face are the main factors related to aesthetics.

To evaluate the harmony and balance of the face, artists, scientists and scholars have established certain norms. These may be taken as a standard for evaluating facial esthetics. Aesthetics and function often go hand-in-hand. Aesthetic facial evaluation is usually done with the person seated in a relaxed and comfortable position. Many of the persons with gross deformities of the face have certain mannerisms related to the facial musculature. Those who have an open bite and incompetent lips try to appose their lips very often. Persons who have vertical excess of the maxilla (upper jaw) with a gummy smile do not smile heartily, and usually try to restrict the upward movement of the lips while smiling.

Harmony and balance are very much related and dependent on facial proportions. Bizygomatic width to facial height is 0.88 for males and 0.86 for females in a well-proportioned face. The ratio of bigonial to bizygomatic width is about 0.70. Likewise, a face that tapers down to the chin is more pleasing.

Face can be roughly classified into three types: Dolichocephalic means a long narrow face with relatively V-shaped dental arches. Brachycephalic means a broad and short face with broad and round dental arches. Mesocephalic faces are in-between the former two categories with a parabolic arch.

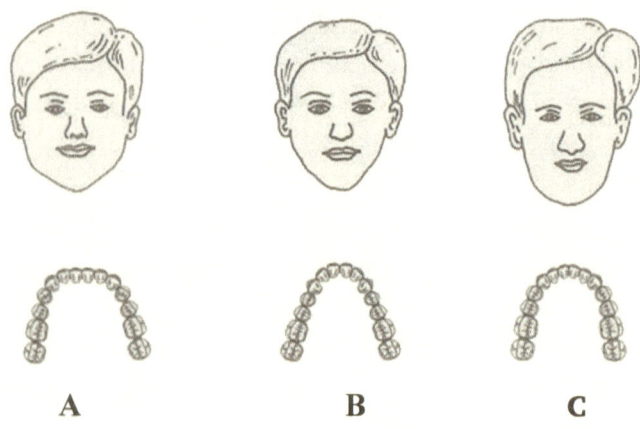

A **B** **C**

Fig 38 *A. Brachycephalic. B Mesocephalic. C. Dolichocephalic.*

In general, a face can be divided into three equal parts from the hairline to the menton (chin). The upper third extends from hairline to glabella (base of the forehead), middle third from glabella to subnasale (where the columella joins the lip), and lower third from subnasale to menton. (The menton is the lower part of the chin. Glabella is where the forehead and the nose meet. Subnasale is where the columella meets the lip.)

Homo Sapiens Divine

Fig 39 *The face can be divided into three almost equal parts from the hairline to the menton; equality of these parts gives harmony to the face*

Upper third of the Face

This extends from the hairline to the glabella. In dentofacial deformities, this part of the face is considered less important due to varying hairstyles. However, in craniofacial syndromes (congenital deformities of the skull and face), analysis of the upper third of the face is important.

Middle third of the Face

This part extends from the glabella to the subnasale. Important structures in this part of the face are the eyes, malar bones, cheek and nose. Intercanthal distance (the distance between the medial part of the eyes) is 34 ± 4 mm and interpupillary (between the pupils) distance is 65 ± 4 mm. These values are usually established at the age of six to eight years and do not vary much. Morphologic variations of the eyes are observed in different races.

Lower third of the face

This extends from the subnasale to the menton (lowest part of the chin) and equals the middle third length. The smile is one of the most important aesthetic factors, and this depends on the tonicity and competence of the lips, contraction of the perioral structures, visibility and angulation of the teeth and dental midline.

Facial aesthetic surgeons follow certain norms to analyze the harmony and balance of the face. The relationship between the structures of the face tells upon facial beauty. Certain salient points are listed below:

Ratio of middle third of the face to lower third of the face is equal to 1:1

Subnasale (root of the columella) to stomion (line separating the upper and lower lip) is half of stomion to menton (lowest portion of the chin), that is, in the ratio 1:2

Subnasale to lower lip vermilion border and lower lip vermilion border to menton are almost equal in length

At rest position, lower lip vermilion is twenty-five percent more exposed than upper lip vermilion

The interlabial distance in normal cases is zero to three millimeters

Interpupillary distance equals the length from commissure to commissure

At rest, the upper incisal border is in line with upper lip border or the teeth are exposed up to three millimeters

While smiling, the upper lip margin lies on the gingival margin

The width of the nose is approximately equal to the inner intercanthal distance

Width of the mandible at the Gonial angles (the angle at the posterior aspect of the mandible) should be approximately the same as the width of the outer border of the orbits

The dental arch midline should be in harmony with other midpoints; the chin should be symmetrical

Normal intercanthal distance is 32 ± 3mm in Caucasians, 35 ± 3 mm in Africans and about 34 ± 3mm in Orientals

Normal inter-pupillary distance is 65 ± 3 mm

Width of the nasal dorsum is one half of the intercanthal distance and the width of the nasal lobule is two third, of the intercanthal distance

Normal upper lip length from subnasale to stomion is 22 ± 2 mm in males and 20 ± 2 mm in females.

The line tangent to the globe of the eye, perpendicular to the Frankfort horizontal plane should fall in the infra orbital soft tissues, ± 2 mm (Frankfort plane is roughly the line drawn from the external auditory meatus through the lower border of the orbit)

The naso-labial angle varies from ninety to 110 degrees; an acute naso-labial angle is an indication of protruded teeth and/ or upper jaw

A line drawn from perpendicular to the horizontal plane through the root of the columella should pass through the upper lip vermilion border and two-millimeter anterior to the lower lip vermilion border a four-millimeter anterior to the most prominent part of the chin. Two millimeters (give or take) is considered to be within normal limits

A straight line drawn from the tip of the nose to the chin (aesthetic line) should pass just in front of the lips

Nasal projection normally should be sixty percent from the base of the columella

Fig 40 *Lower third of the face—from subnasale to stomion is half that of stomion to menton*

Fig 41 *The naso-labial angle varies from ninety to 110 degrees; acute angle is an indication of protrusion of the upper teeth if the angulation of the columella is normal*

Fig 42 *Labio mental sulcus is about four millimeters deep*

Fig 43 *Subnasale Perpendicular to F H plane passes through upper lip vermilion and two-millimeter anterior to lower lip vermilion and four-millimeter anterior to the chin. Though simple this is an excellent method to assess the relationship of the facial features. Nasal projection angle formed by a tangent on the dorsum of the nose and the perpendicular to the FH plane is normally thirty-five degrees, ± 2 degrees*

Fig 44 *Aesthetic line*

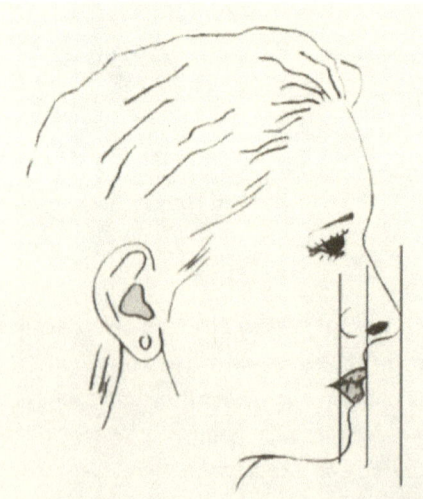

Fig 45 *Drawing perpendicular lines through the nasal tip, upper lip vermilion border, and the alar cheek junction can assess nasal projection, which is about sixty percent*

Animation gives a face its character. It is difficult, if not impossible, to correct learned habits and expressions of facial muscles. So, evaluation of a relaxed smile, full smile and so on will help assess symmetry of muscle movement, action and hyperactivity of muscle groups.

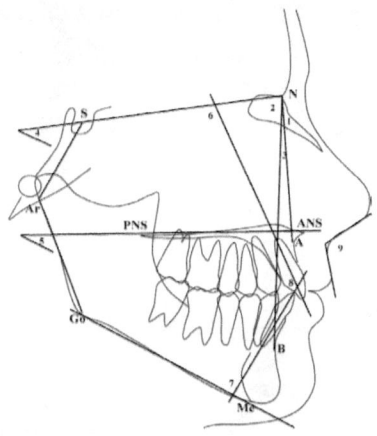

Fig 46 *Cephalometric analysis is a useful aid in assessing the skeletal harmony, balance and to pinpoint the deformity for maxillofacial surgeons to plan the surgery*

McNeil, Proffit and White were the first to discuss the use of cephalometric prediction for orthognathic surgery patients. In 1931, Broadbent developed the new X-ray technique and described its application in cephalometrics. Later, many investigators proposed cephalometric analysis for orthodontics and orthognathic surgery, which included soft tissue and skeletal analysis. In Steiner's analysis, angles SNA (S is Sella, N is Nasion, A is deepest point of the maxilla in the anterior part below the anterior nasal spine) and SNB are measured to evaluate the relative position of maxilla and mandible to the cranial base. It is one of the most popular methods. In the McNamara analysis, maxillary position is evaluated in relation to a Nasion perpendicular to anatomical Frankort Horizontal line. Point A is on this line or slightly ahead.

In a normal face, the sella–nasion (SN) plane is taken as the reference plane. ('S' is the central point of sella tursica, which is the seat of pituitary gland). The Frankfort plane is about seven degrees plus or minus one to the SN plane. The palatal plane will be about eight degrees plus or minus three, the occlusal plane will be fourteen degrees plus or minus four and the mandibular plane will be thirty-two degrees plus or minus three to the SN plane. All these planes, if drawn backward, will meet at a point behind the cranium.

Homo sapiens Divine

Physical and intellectual changes took place gradually and consistently during the evolution of hominids. These changes were for the betterment and perfection of the species. Nature is very conservative and economical. Organs that are of no more use due to various reasons lose their function and gradually become vestigial, and even disappear by natural selection. Depending on the demand, new body parts may emerge and take up new functions. As the importance of evolution shifted from physical to intellectual, the brain gained more importance.

The striking difference between humans and their closest living relatives (chimpanzees and gorillas) is in the size of the brain and their lifespan. The brain is larger by a factor of three to four, and the lifespan is larger by a factor of two. According to Kettard S. Kaplan and Arthur J. Robson, the larger human brain is an investment with initial costs and subsequent rewards.

The cranium of the modern man is characterized by its globularity and facial retraction. According to Daniel E. Lieberman and associates, the developmental changes that led to the modern cranial form were derived from a combination of shifts in the cranial base angle, cranial fossa length and width, and facial length.

Hominins were the first animals to 'grow' an extended limb outside of them by making tools. Tools are considered as extra corporeal limbs. Prof John Crawford is of the opinion that this has reduced the pressure on the body for frequent adaptations. However, this increased the stress on the neural tissue to develop intelligence to use and modify the tools. Absolute brain volume has more than tripled from Australopithecus afarensis (480 cc) to Homo sapiens (1500 cc) and the relative brain size has more than doubled.

Full appreciation of objects and events in the external world was dependent upon the development of the brain cortex. According to Elliot Smith, this occurred only when man became human. Analysis of pre-human and modern human skulls shows a tendency for increase in frontal bossing during evolution. This is due to the increase in the size of the frontal lobe, which is considered to be the seat of intelligence. Thus, the slanting forehead became straight.

W.P. Rock et al conducted an interesting study in 2006. They compared human skulls from the 14th, 16th and 20th centuries. They found that the horizontal measurements in the anterior cranial fossa and the maxillary complex were greater in the modern group. The cranial vault, especially anterior cranial fossa, was significantly higher in the modern skull. They concluded that the medieval ancestors had more prominent faces and smaller cranial vaults than the later group. The angle SNA was lesser in the modern man, indicating a reduction in the size of the maxilla. The prefrontal area is accepted to be the seat of intelligence. Increased intra-cranial dimension and high forehead of the modern man are evidences of the increase in brain size over centuries.

This evidence is good enough to show that evolution is occurring at a much faster pace than generally thought. The brain size in relation to basicranial length is an important determinant of the basicranial angulations. Spoor et al (1999) and Liebermann (2000) opined that differences in cranial base angle are more likely to account for facial retraction in modern humans. Despite the lack of fossil proof, biologists of early days, who were proponents of evolution, such as Lamarck, Huxley and Darwin, speculated that bipedalism preceded encephalisation (enlargement of the cranium and brain). Their speculation was proved right later by archeological studies and fossil findings. The frontal lobe and cerebrum are supposed to be the seat of intelligence. One of the striking features that delineate Homo sapiens (the wise man) from his ancestors is the prominence of the forehead by its steep rise. The modern human skull looks infantile when compared to the neanderthal skull with the cranium more round and delicate. Comparison of the brains of amphibians, reptiles, birds and mammals shows that the increase of the brain size is mainly due to the enlargement of the cerebrum.

Facial Changes

As humans climbed the ladder of evolution,
Jaws lost many a function;
Size suffered a gradual reduction;
The brain enlarged for better cognition;
The face moors the fine human expression;
In the evolutionary bargain;
A unique smile was a charming gain.

The reduction in the size of jaws and teeth could also be attributed to the development of tools and transfer of functions from the jaws to the forelimbs. In fishes, even the breathing function is by the jaws. In mammals, the jaws perform many functions such as carrying cubs and prey, and fighting, besides chewing food. The modification of the skull was a gradual adaptation and evolution when the arboreal ape climbed down from the tree and had to travel through the terrain. In the terrain, he had to travel fast to escape the predators as well as to reach the prey. Bipedal locomotion is superior to knuckle walking when the need is to move fast. For upright locomotion, a protruding jaw was a hindrance as it blocked the vision of the immediate ground in front—this necessitated a reduction in size of the jaw. The released forelimbs took over many of the jaws' functions. For better dexterity and skillful use of the hands, stereoscopic vision became necessary and the eyes were pushed from the sides to the front. To support the skills, cortical mechanism developed. Smith opined that as development of the tools (extra corporeal limbs) occurred, the demand on physical evolution reduced and a demand on intellectual evolution increased—thus resulting in increased volume of the brain.

Certain other morphological adaptations occurred along with bipedalism. These include the valgus knee angle, anteriorly placed foramen magnum, and short, broad, bowl-shaped pelvis. As the jaw size reduced, the masticatory muscles also reduced in size. For the ape, the jaw is broadest at the canine area. However, in man, it is broadest at the condyle region. This is because the skull has expanded in lieu of the enlarged brain, pushing the glenoid fossa laterally.

During embryonic development, the mandible develops as two halves, which are joined in the midline. This area is subjected to great strain, owing to the powerful muscles of the apes and pre-humans. To contain the stress at the inner side of the anterior part of the jaw, a shelf of bone called the 'simian shelf' developed. In humans, the muscle size reduced, jaw size reduced and the simian shelf disappeared. A remnant of the shelf is the genial tubercles to which the genioglossus and geniohyoid muscles are attached. The genioglossus is an important muscle in the movement of the tongue. The chin moved forward and the jaw was widened at the posterior region to give more space on the inner side, providing more maneuverability to the tongue that helped in the development of speech. Paul Broca, a great French anthropologist, discovered that the third inferior frontal convolution is the speech area—which was later named after him. Now, it is understood that this is only one of the cortical areas associated with speech.

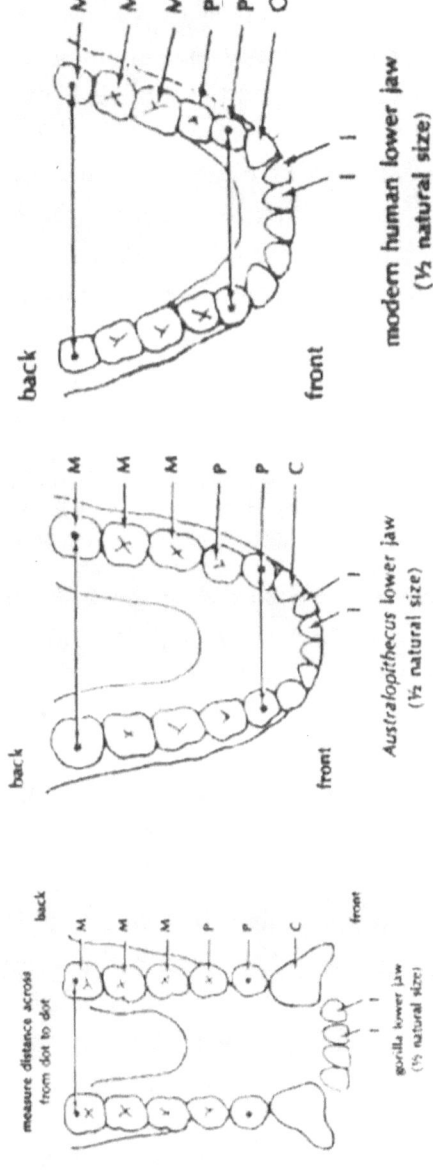

Fig 47 *The lower jaw was widened at the condylar region as the skull enlarged*

Among the Hominins, Homo sapiens are the only species that have a well-developed chin. A protruding chin is vulnerable to injury in a fight or accident. In response to this, the neck of the condyle became thinner and the sigmoid notch became deeper. Use of direct force on the chin will be transmitted to the condyle. If the condyle's neck is broad and strong, the force could fracture the temporal bone and, in turn, injure the brain, thus killing the victim. Hence, a thin condylar neck was a favorable evolution for protection of the brain.

According to Rebecca Rogus Ackermann, genetic drift should be the primary cause for facial diversification. However, selection might have also played a role in the changes in facial morphology during the evolutionary process.

Reduction in the size of the jaws during human evolution is associated with its shift under the skull. This had an influence on human dentition. A beautiful regular smile is the result of a delicate balance among many different genes that determines the facial features. So, a regular smile, sign of good genes, some theorists claim, is the underlying sexual significance of physical beauty.

The face, jaws and teeth of Mesolithic humans of 10,000 years ago were about ten percent more robust than those of modern humans. Thirty thousand years ago (Paleolithic period), it was about twenty to thirty percent more robust. Relative to the body size the, molars of Australopithecus afarensis were 1.7 times larger when compared to that of modern species of hominins. An interesting finding is that the size of the teeth has increased in Australopithecus robustus, and further evolution decreased the size of the teeth. This could be attributed to food processing habits (using tools), which the later forms developed.

Orbit and Eye

Eyes owe their beginning to the sensitivity of protoplasm to injuries and the beneficial effects of light. Early pre-vertebrates had directional organs that used light. True vision resulted as the brain enlarged and accessory organs developed to change the curvature of the lens for focusing light. The orbits moved forward and stereoscopic vision developed. Chimpanzees and humans have chromatic and non-chromatic vision. As the face reduced in length, the size of the orbit also reduced in height.

In earlier primates and Hominins, the brow ridge is very large—like a penthouse protecting the eye. These large brow ridges act as buttresses to resist the stress exerted by the massive jaws while chewing. As the jaws and the muscles of mastication became smaller, the parietal bone became prominent. The brow ridge lost its function and got reduced in size.

Lips

A leathery skin that could be a remnant of a tough reptilian skin surrounds the mouth of the 'duckbill', an archaic mammal. In the spiny anteater, the lip has muscles and is covered by hair. As evolution progressed to the great apes, the lips were protruded. The philtrum of the upper lip is peculiar to the Hominins. As the jaws shrunk in size, the alveolus was also reduced and the lips rolled outwards, exposing the mucosa-lined part of the lip. Another theory suggests that the protracted period of suckling could be the reason for eversion of the lip.

Nose

Jacobson's organ, found in amphibians, reptiles and primitive mammals, allows the food taken through the mouth to be smelt. It is absent or vestigial in the adults of higher primates and man, but is found in foetal life. Among the primates, the modern man's nose is made prominent by the elevation of the bridge and prolongation of the tip.

Reduction of the lateral part of maxilla and the movement of orbits medially might have narrowed, elevated and arched the nasal bone, forming the bridge of the nose. The tip of the nose is not vestigial, but evolving. The shape of the nose in man is mainly due to the regression of the jaws. The rising bridge of the nose might have helped to give resonance to man's voice.

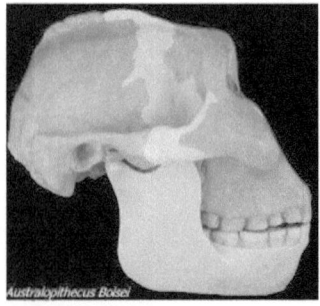 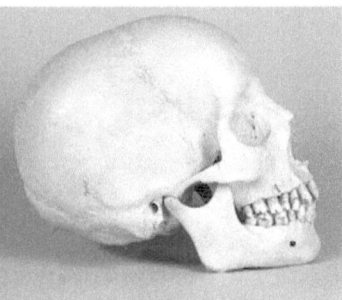

Fig 48 *Comparison of the skulls of Australopithecus bosei and Homo sapiens*

Summary of Evolutionary Facial Changes in Humans

Brain case enlarged and become rotund

Frontal bossing has taken place making the forehead straight and higher

Nose became prominent

Chin became prominent

Size of jaws and teeth were reduced

Muzzle became shorter

Teeth became less protruded

Diastema (space between the teeth) reduced

Teeth became smaller

Simian shelf reduced to genial tubercles, giving more maneuverability for the tongue

Last molar teeth (wisdom teeth) have become vestigial or absent

Nasal bone became prominent

Overall size of the jaw got reduced

As the skull got enlarged, the temporal bones were pushed apart, making the human jaw wider at the posterior region

Sigmoid notch has become deeper

Neck of the condyle became thinner

Brow ridge reduced in size

Face became more oval

Neoteny has set in

The concept of beauty varies from time to time, place to place and race to race. However, there are certain inherent norms and universal concepts about beauty. A multitude of ideas about beauty is evident from the host of statements by many authors, philosophers and artists. A statement by Shakespeare, 'Beauty lies in the eyes of the beholder' (*Merchant of Venice*), suggests that beauty is subjective.

'A thing of beauty is a joy forever

Its loveliness increases; it will never

Pass into nothingness.'

(John Keats, *Book 1 of Endymion*)

These lines depict the positive influence of beauty on emotional perception. Keats glorifies beauty. True beauty radiates vitality and happiness.

Nancy Etcoff writes in her book, *Survival of the Prettiest*: "In the animal world gaudy plumage and body ornaments, emerge at sexual maturity and animals reserve their brightest colors for courting displays. Caterpillars turn into butterflies and peachicks explode into psychedelic colors of the peacock, when it is time to reproduce. Flowers are alluring landing strips for pollinating insects: They are the plant world's sex objects. Throughout the natural world beauty is the harbinger of sexual reproduction."

Most of the white flowers like jasmine are fragrant and are pollinated by nocturnal insects. Orchids mimic the shape of the female wasp, and the male wasp tries to copulate with it—thus unknowingly helping in pollination. Some flowers reward the insects with sweet nectar for pollinating them.

Opinions abound about the beauty of the face, as it is more subjective than objective. Some consider attractiveness as having average ingredients. However, Francis Bacon, in his 'Essays on Beauty', opines: "There is no excellent beauty that hath not some strangeness in the proportions." However, symmetry has a positive influence on facial beauty *(Grammer K., Thornhill R.)*. Attractiveness is related to positive physical qualities.

Organisms could survive by challenging the odds or by adapting to the periodic positive and negative changes taking place in nature. Fitness is often a harmonious and balanced blend of both. Meekly getting adapted to the surroundings to exist doesn't mean progress—if so, the thrust of evolution is not to advance but to compromise, and the result will be just survival or passive existence. In nature, we find not passive existence but a positive progression of life that empowers the species to face the challenges and odds. Hence, 'fit to survive' takes on another dimension 'fit to conquer', strengthening the species to progress through evolution. All-round progression has taken place in metabolism, reproduction, form and function.

Anaerobic metabolism has been replaced by aerobic metabolism, and it is more efficient in the production of energy. This happened due to the change to an oxygen-rich atmosphere. One molecule of glucose can produce two ATPs (energy packets) in anaerobic metabolism, but the same molecule can produce eight ATPs by aerobic metabolism. As evolution progressed, organs became specialized and the regenerative capacity of the organism regressed.

A single-celled organism never dies by ageing, but multiplies by binary fission. It dies only by injury or starvation. The next step in the evolution of reproduction was conjugation and division, where genes were allowed to mix. A single-celled organism, by evolution, became a multicelled organism with functional differentiation of parts, and had to accept death by ageing. In lower forms, each cell is pluri-potential, which has the property to grow into a full organism. Vegetative reproduction in plants is an extension of this phenomenon. Gradual evolution has seen sexual differentiation and sexual reproduction. This has improved and increased the pace of evolution by mixing of genes and promoting selection.

Sexual reproduction paved the way for the selection of a partner. Purpose of evolution, it appears, is to attain perfection, and, hence, it is important that the organism moves on to better fitness in each generation, and to a better species. During the progression of evolution, the stress was shifted from physical to intellectual evolution by gradual refinement of the nervous system to a well-formed controlling brain. This profound shift might have happened in a reptile (Sagan Carl, *Dragons of Eden*).

(In the story of Genesis in the *Bible*, the fruit of knowledge was given to Eve by a reptile.)

(Comparison of bits of information in the genes and in the brain presents a very interesting observation. For amphibians, the information in the gene is much more than the information in the brain, but it is the opposite in mammals and maximum in humans. If we observe the graph, we find that the transition took place in a reptile. We can observe that, from then on, the focus of evolutionary thrust was more on the brain than on the body.)

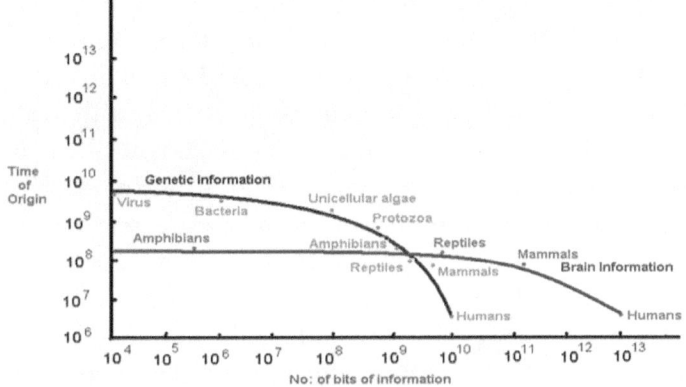

Fig 49 *(Idea taken and modified from 'Dragons of Eden' by Carl Sagan)*

Sexual reproduction brought out the need for the selection of a partner for reproduction. Attractiveness became an important criterion for selecting a partner. External features representing health, vitality, and intelligence, by default, became attractive. Given the cascade of rejections and selections during the evolutionary process, attractive features were incorporated into fitness and the 'survival of the fittest' became akin to 'survival of the prettiest.' True beauty evokes pleasure and pleasure is perceived in the subconscious limbic system, not in the cognitive neocortex.

Many studies have shown that a baby's facial characteristics, such as large and round eyes, larger domed forehead and well-defined chin are attractive features, especially on females. It is to be noted that, during the hominin evolution, tendency towards retention of childhood characteristics was evident. This tendency is called 'neoteny.' Humans exhibit a number of neotenies compared to apes. In chimpanzees, adulthood starts at around two to three years. In humans, it is at fourteen. Some of the features of the face that are attractive are brown skin, full lips, larger distance between the eyes, larger and darker eyelashes, darker eyebrows, high cheekbones, narrow nose and so on. People with more attractive features were assessed to be more successful, content, pleasant, intelligent, sociable, exciting, creative and diligent. This finding poses a question as to whether intelligence and other positive qualities cultivate attractiveness, or if the confidence imparted by attractiveness in turn makes people more intelligent and creative. In fact, these are interdependent and contributory. If attractiveness and intelligence are extrapolated in time, we can deduct that both are the main factors in the selection of a mate—survival of the prettiest and the brightest or, in other words, survival of the fittest. This plays an important role in evolution.

Leonardo Da Vinci drew the 'Vitruvian man', relating divine proportion to the human anatomy. The body, when in divine proportion, is able to effect maximum efficiency with least effort and also impart better aesthetics, reflect health and provide survival advantages on the aspect of selection and propagation of genes and subsequent evolution, function always precedes form. Hence, it can be postulated that prettiness is a manifestation of fitness and brightness.

Attractive physical features provide external clues to health and fertility status—the two most important requirements for genetic success. "In civilized life, man is largely influenced in the choice of his wife by external appearance"—Charles Darwin noted so, way back in 1871. Longitudinal data suggest that attractive women tend to marry men in high occupational positions. Henry Kissinger once said, "Power is an aphrodisiac." Testosterone, a male hormone, makes the person grow big and also causes the lower jaw to grow longer. In the world of animals, big territories and power are attractive features used by males to lure females.

Estrogen, a female hormone, increases the lip volume and fat deposits on the buttocks and thighs, and increases the utilization of abdominal fat, thus giving the woman an hour-glass (gynoid) appearance that is associated with menarche. Large breasts give the impression that there is plenty of milk for the child, which is important for rearing a healthy offspring—a fundamental prelude to propagation. Wide girdle bones can house a large head of the foetus, which means a large brain volume (more intelligence). Humans are considered to be the species that undergo the maximum pain during labor, and have the highest maternal mortality rate. This could be due to the fast intellectual evolution (large head of the foetus) and not so fast physical evolution (pelvic bone). (For taking the forbidden fruit of knowledge, woman was cursed by God that she would deliver her children in pain—*Genesis, Old Testament, the Bible*.)

Humans have an innate capacity for appreciating balance and harmony. As the face is the prime seat of expression, it is the area of attention. Study of the evolution of the face and skull gives insight into the evolution of facial beauty.

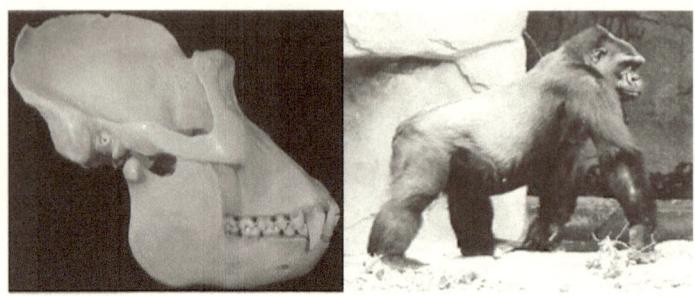

Gorilla

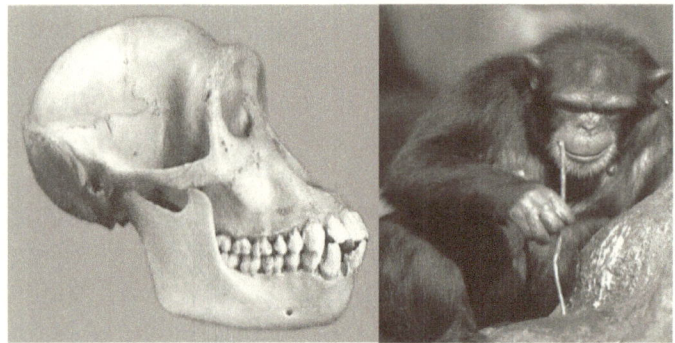

Chimpanzee

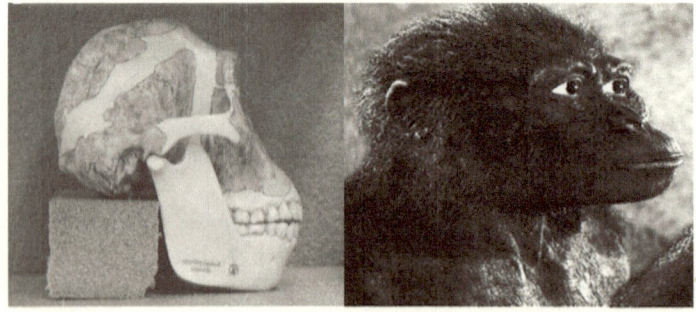

Australopithecus africanus

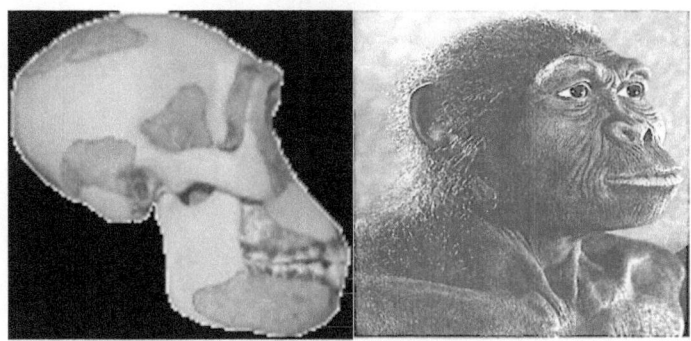

Homo habilis, the skillful man

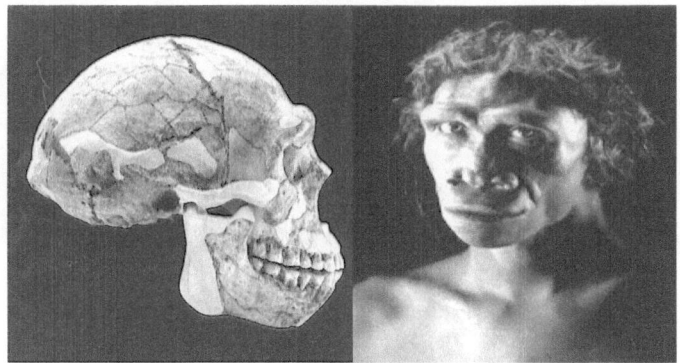

Homo erectus, the erect man

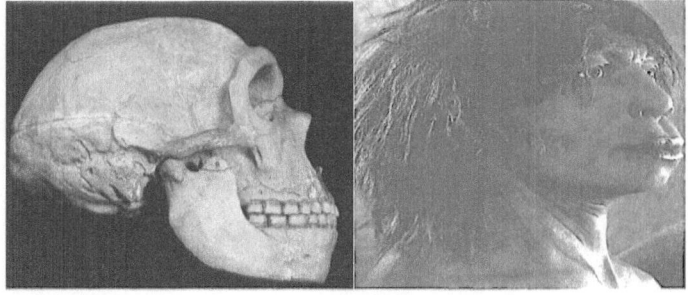

Homo neanderthalensis

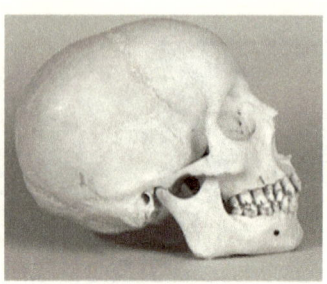

Homo sapiens

Fig 50 *Comparison of the shape of the skull and the face of different species of our ancestors. We can appreciate the enlargement of the cranium and reduction in the size of the face*

Cephalometric analysis to assess and compare the following parameters of different species of the Hominidae family: 1) Maxillary projection (angle MNI) 2) Mandibular projection (angle MNB) 3) Frontal bossing (angle FB) 4) Cranium-facial proportion in two planes—frontal and lateral. To analyze beauty and the proportion of expansion of brain case and reduction of the jaw size, a modified cephalometric analysis has been applied here.

MN Line: MN line is drawn from the junction of the frontal bone and the nasal bone (fronto nasal junction—N) to the base of the mastoid process (point 'M'). This line virtually separates the brain case from the jaws.

Point 'I': Tip of the upper incisor is taken as point 'I.'

Point 'B': A parallel line to the occlusal plane is drawn at the middle point from occlusal level to lower border of the mandible at the first molar region. The point where the line meets the anterior border of mandible in the lateral profile of the skull is taken as point 'B.'

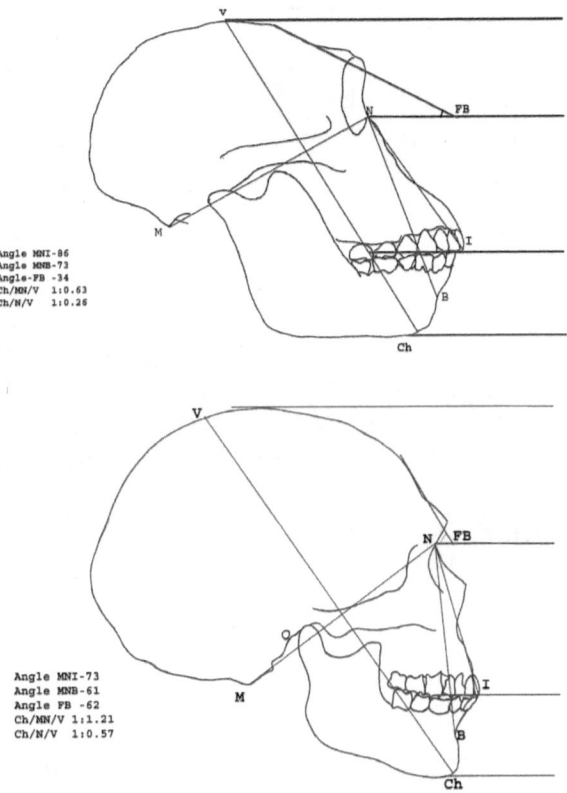

Fig 51 *Cephalometric analysis of Australopithecus africanus and Homo sapiens*

Point 'Ch': Point where anterior border and lower border of mandible meet is marked as point 'Ch.'

Point 'V': A perpendicular line to 'MN' line is drawn from point Ch to the top of the skull. This point is marked, as Point 'V', to mark the vertex.

Angle MNI: Gives projection of the maxilla. This is found to be decreasing from the Australopithecus afarensis to Homo sapiens in the order of evolution. This indicates the reduction of maxillary size.

Angle MNB: Angle MNB gives the anterior projection of the mandible. It is also found that this angle is decreasing and it indicates that the antero-posterior length of mandible is decreasing.

Angle of frontal bossing: This is calculated by drawing a tangent of the frontal bone at the anterior plane. A horizontal line parallel to the occlusal plane is drawn from point N. Angle formed between these two lines is the angle of frontal bossing. It is found that this is increasing during the evolutionary process due to the increasing size of the brain, especially the frontal lobe.

Skull–face proportion:

When we discuss the proportions for beauty it is imperative, that we consider phi (ϕ), the golden proportion, which is 1:1.618.

'When a line is divided in such a way that the ratio of the shorter section to the larger section is equal to the ratio of the larger section to the whole line, this is supposed to be the most aesthetically pleasing'—this is known as golden proportion and is represented by the symbol ϕ (phi). The name 'phi' is derived from the Greek sculptor Phidias, who used golden proportion in his most famous work, Pantheon. This proportion is linked to so many aspects of beauty that Kepler called it 'divine proportion.'

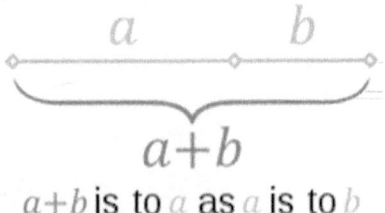

$a+b$ is to a as a is to b

Fig 52 $A+B=C$; $A/B = C/A = 1.618$

$A/C = B/A = 0.618$.

Phi is supposed to be the most pleasing proportion and was used in the construction of Greek temples and by Michelangelo and Da Vinci in their paintings. Michelangelo's famous painting, 'Temptation and Expulsion from Eden', is a well-known example of this proportion. It appears that, in nature, this system of designing is abundantly observed in living and even non-living structures. The design of leaves, petals of flowers, shells and so on follow a pattern called the 'Fibonacci series.' (0,1,1,2,3,5,8,13,21,.......), wherein the preceding two numbers are added to get the third number. It is interesting to note that the proportion between the adjacent two numbers is about 1.618, which is known as the golden or divine proportion. (Example: 8/5= 1.6; 13/8= 1.62; 21/13=1.61 and so on)

The proportion between the cranium and the face is calculated on two different planes—frontal and lateral. The distances between vertex (V) and the nasion (N), and from the nasion (N) to the lower border of the mandible are taken in a vertical plane. This proportion in the frontal plane appears to approach the golden proportion as the evolution progresses—(Table 1) and (Fig 51).

Fig 53 *Nautilus. The chambers are arranged in a spiral format with size of the chamber increasing by the golden proportion. The Fibonacci series and phi proportion are abundantly seen in nature*

Table 1 *Over the last three million years, about seven important Hominin species evolved. During this evolutionary process, the maxillary and mandibular projection has decreased. However, the frontal bossing has increased. The cranium–face proportion has increased in favor of the cranium, which enlarged and became globular. (MYA—Million years ago; MNI—Projection of maxilla; MNB—Projection of mandible; FB—Frontal bossing; Ch/MN/V—Proportion of the dimension cranium and the face on the lateral plane; Ch/N/V—Proportion of cranium and face on the frontal plane.)*

Species	Years (MYA)	Angle MNI	Angle MNB	Angle FB	Ch/MN/V	Ch/N/V	Brain Vol. (cc)
Australopithecus africanus	3	86	73	34	0.63	0.26	420–500
Australopithecus bosei	2.5	80	70	26	0.7	0.27	480
Homo habilis	2	78	69	51	0.92	0.5	650
Homo ergaster	1.5	76	65	43	1.06	0.5	850
Homo erectus	1	75	65	46	1.06	0.5	800–1000
Homo neanderthalensis	0.5	74	65	50	1.05	0.51	1400–1800
Homo sapiens	0	73	61	62	1.21	0.57	1040–1600

On the lateral aspect, the MN plane virtually divides the cranium and the face. As the brain case is expanding and the jaw size is reducing, this proportion will approximate the divine proportion as evolution progresses. If the extrapolated graph (of the proportions from the pre-human to human face through different species) is projected into the future, we can deduce that it will take another million years for the present proportion of skull and face to reach divine proportion.

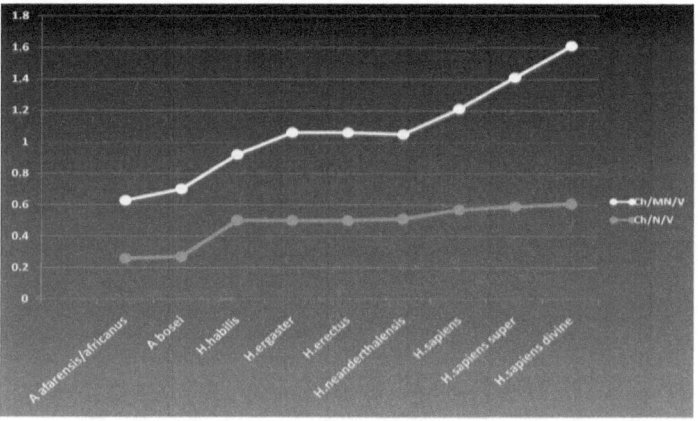

Fig 54 *This graph of the proportion between the cranium and the face is projected in time to reach the golden proportion. To reach the golden proportion, it may take another one million years and two more advanced species*

If we go through the archeological history of the Hominidae, branching of the humans took place about three million years ago, and, from then on, about eight important species emerged to reach the present Homo sapiens. It is also to be noted that during the first three to four million years from the branching of the hominid from his cousins, the apes, evolution was slow and became faster pace during the latter half. So, we can expect two more species of humans evolving during the next one million years. By then, the adult human face may look very similar to that of the present child, with a large head, large eyes, small jaws and an oval face. Children's faces often exhibit the divine proportion.

The superman concept of Nietzsche and the statement, "Man is poised between the beasts and the gods" by Plotinus, points to the evolution of Homo sapiens to a more beautiful and intelligent being. Can we call the future two spices 'Homo sapiens super' and 'Homo sapiens divine' respectively?

Fig 55 *The future human may look like this 'Barbie doll' with a large head, large eyes, small jaws and oval face*

9
Culture and Existence

Cultural evolution can be related to the development of the nervous system. Important requisites for any living organism are self-preservation, reproduction and other biological needs. As the nervous system progressed to the brain through the addition of certain parts for better adaptation and progress, the needs of the organism evolved from biological to self-actualization in humans. These needs are the foundation of cultural evolution. Not only humans but many other animals became social creatures as well. Our culture evolved through millions of years, influenced by our behavior and reasoning capacity.

Our early ancestors might have behaved like chimpanzees. We share 98.4 percent of our genes with them. Chimpanzees do not gather food—they eat when they find it, then and there. However, when they find meat, others in the group ask for a share. Proto-humans might have started sharing not only meat, but plant-based food as well. Gradually, group hunting and gathering of food developed, and started bringing them to a central area for sharing. Chimpanzees lived in the tropical forests, where vegetables and fruit were plentiful—when food is abundant, there is no need for sharing or hoarding. Proto-humans lived in the fringes of the forest in the Savanna grassland. Food was not easily available. This created evolutionary as well as cultural pressure on the early humans. Group hunting, gathering, storing and sharing developed as a response to the environmental pressures of scarcity. Division of labor emerged as a result. Sexual dimorphism was a striking characteristic of the early humans—so, naturally, more risky jobs that demanded physical strength, such as hunting, might have been taken over by males, and the collection of plant-based food and looking after the young ones were females' tasks. Initially, foraging might have been scavenging—collecting meat from the animals killed by predatory animals. Females are more important biologically as they are the ones who conceive and bear the child. It would be deleterious for the species to lose a female than losing multiple males. Biologically, the job of a male is to impregnate the female—it is unreasonable to put a fertile female in the risky job of hunting, as the predator could become prey at any moment.

Grouping provided great many advantages—the most important was protection, followed by sharing of food. Humans depended on meat, brought by males, and vegetables gathered by females. More than reproductive cooperation, social cooperation and mutual dependence developed between both the sexes. The group had to travel and camp for foraging. Migration of the species thus occurred, and promoted the development of new characteristics and even speciation. With migration, humans have become a global species. They reached the Arctic 37,000 years before present (BP), went to Europe about 45,000 years BP, Australia 50,000 years BP, China 60,000 years BP, South Asia 70,000 years

BP. Since the Pacific Ocean was a challenge, humans occupied New Zealand only about 1,000 years ago. Globalization improved their exploration capacity, and their ability to adapt to the new surroundings. The pre-human ancestors became extinct because of their lack of capacity to explore and migrate or adapt to the changing global environment.

Women are more group-oriented, have sharper senses, and can read faces better than men *(Moir, 1989; Tannen, 1991)*. Females are vested with the task of rearing the young ones. For humans, this is important as they have the longest childhood and need training from the very beginning. Mothers impart the initial training. Most of the fishes and reptiles are born with the capability to survive on their own. Mice are ready in months, canines in weeks, lions and wolves in months. Chimpanzees take around two to three years to be on their own. Humans take around twelve to fourteen years and needs specialized training to face the world. The early humans may fall between the chimpanzees and the modern man in this regard.

The development of tools was another important episode in the cultural evolution. This was associated with the evolution of an opposable thumb. The beginning of tool-making is attributed to the Homo habilis (skillful man). Tool-making gradually developed into throwing stones, javelins and to bow and arrows. The development of language and abstract thought was a quantum leap in cultural evolution with definitive cultural evolution.

The first Hominids evolved in Africa about four-and-a-half to three-and-a-half million years ago. They were bipedals with larger brains than the contemporary apes. The earliest stone tools were found in Ethiopia around two-and-a-half million years ago. The Homo habilis are considered to be the ones who started making stone tools and hand axes. About one-and-a-half million years ago, the Homo erectus emerged and they had the capacity to communicate by speech, though it might not have been syntactic. Around this time, the Australopithecus became extinct. About one million years ago, the Homo erectus started migrating to Europe and Asia. Latest fossil finds of Homo erectus were only 30,000 years old. About 130,000 years ago, fire was used by the Homo erectus. Archaic humans started constructing huts from animal bones and wood. They buried their dead with rituals, which means they either believed in supernatural powers or life after death. This may be considered as a higher level of abstract thinking. Neanderthals were living around 130,000 to about 35,000 years BP in Europe and Asia.

The cro-magnon is considered the first modern human—Homo sapiens sapiens. They appeared about 90,000 years ago and replaced neanderthals in Europe. They used bone tools and had a reasonably well developed language. Paintings in the cave walls of Chauvet caves at Vallon-Pont-d'Arc, France, are attributed to the cro-magnon. A sculpture that was about 28,000 years old was found in Austria, and this was named 'Venus of Willendorf.'

Domestication of animals started around 14,000 years ago, in China. The first ones to be domesticated were dogs—they were good companions for the humans in hunting and for security. Domestication of sheep,

Culture and Existence

goat and horses came later. Agriculture started only about 10,000 years ago. This is to be considered as an important cultural evolution—man has started settling down instead of being nomadic.

The invention of the wheel was the beginning of transport. Invention of numbers (digital system) and written language developed in Egypt around this time. The invention of papyrus (paper) also happened in Egypt. The oldest Kingdom was also in Egypt—around 5,100 years ago. The oldest pyramid is about 4,700 years old. The Shang dynasty in China is about 3,800 years old.

The first Olympic Games was hosted in Greece 2,800 years ago. By around 2,600 years ago, cultural evolution was at its peak in Greece. Democracy, considered the top social system, developed there. (Bertrand Russell quipped, "Envy is the basis of democracy.") They had well-established laws; art has progressed, and poetry, drama and philosophy flourished. The Roman Empire rose to power by 2,000 BP. During the same period, some people started following the teachings of Jesus Christ. During the initial periods, Roman rulers persecuted Christians. They later embraced Christianity, which spread all around the world.

The Industrial Revolution started around 200 years ago, with textiles, iron and steel. The last century has seen an unbridled progress in science and technology, which has had a profound influence in every field of life and culture. Darwin's theories of evolution (1859), the invention of DNA (1953), genetic engineering (1987), cloning (1996), and the genome project (2001) have revolutionized the biological and medical sciences.

Such revolutions have taken place in every field of life. The invention of the telephone by Bell (1876), electric bulb by Thomas Edison (1879), invention of the television (1930) and so on, followed by the invention of computers and advances in information technology, has had a stupendous influence on our way of life.

However, social, cultural, ethnical, and religious groupings have created havoc in the world. World War I (1914 to 1918), in which thirty-two countries took part, resulted in the death of about twenty million people. In 1917, the Russian revolution took place and the United Soviet Socialist Republics (USSR) came under communist rule. World War II (1939 to1945) saw the death of thirty-five million people. The atomic bomb, used in this war, killed 130,000 people.

Space research also saw soaring progress. The first space voyage was by Uri Gagarin (USSR) on 4 April, 1961. The first astronaut landed on the moon on July 20, 1969. The first robot reached the moon on 4 July, 1997; the first man-made object, Voyager, escaped the solar system to enter interstellar space on 25 August, 2012.

Racial, Ethnic and Religious Discrimination and Conflicts:

Racism was a belief based on the biological differences of people of different ethnicities. It has led to the attitude that some ethnic groups are superior to others in intellect and other qualities. These concepts begot many atrocities such as the trans-Atlantic slave trade, apartheid in South Africa, the holocaust, and global colonization.

In the 19th century, many scientists and philosophers believed that there were different races of humans and that they had separate origins—hence, racial superiority of some ethnic groups was justified. This was known as 'scientific racism.' In 1975, Johann Bluemenbach advocated polygenism, which proposed that different races have different origins. Christopher Meiners, a polygenist, divided humans into two divisions— 'beautiful white race' and 'ugly black race.' He considered the black race to be inferior, immoral and animal-like.

A study conducted by Devah Pager (professor of sociology Harvard Kennedy School) et al in 2009 showed that job preferences favor whites by more than fifty percent than blacks, even if they are equally qualified. This evidently shows that racial discrimination is still prevalent, even in advanced societies in spite of the scientific evidences that show that all humans are of the same species, and that phenotype differences are mainly due to adaptations to environmental variations.

In his book, *The Descent of Man*, Charles Darwin opposed the idea of racial differences and argued that all humans are of the same species with same mental faculties though there is cultural difference between the 'lowest savages' and the Europeans.

According to the gene-centered study and human genome project, where the complete human DNA mapping was done, there is no evidence for genetic basis to differentiate racial groups.

However, there is a tendency among humans to group together on ethnic basis. This could be attributed to inherent xenophobia and kin selection. Richard Dawkins, in his book, *The Selfish Gene*, writes: "Blood-feuds and inter-clan warfare are easily interpretable as an irrational 'kin selection' tendency to identify with individuals having similar physical resemblance."

From antiquity, dark-skinned people were considered inferior in almost all aspects. According to the *Book of Genesis*, Noah cursed his son, Ham, to become the servants of servants. Africans are considered the descendants of Ham, and the slave traders used this story as justification.

During the 19th century, racism was tied to nationalism, which resulted in pan-Germanism, Zionism, and so on. Moreover, most philosophers and eminent men considered the blacks as inferior to the whites. Rudyard Kipling wrote a poem, *The White Man's Burden*, to this effect. The worst of the racial violence was the holocaust, where more than six million Jews and about five million other ethnic non-Europeans were killed by the Nazi regime. The Durban Riot between Indians and Zulus in South Africa in 1949, the exodus of Burmese Indians in 1962, the Zanzibar revolution in 1964, Apartheid in South Africa and so on are evidence of the prevalence of racism in the 20th and 21st centuries.

Ethnic conflict is prevalent in many parts of the world even today. Recently, it appears that racial strife and ethnic conflicts have given way to religious wars, in the name of 'God.'

The International Convention on 'Elimination of All Forms of Racial Discrimination', adopted by the United Nations general assembly in 1965, stated: "Any doctrine of superiority based on racial differentiation is scientifically false, morally condemnable, socially unjust and dangerous, and that there is no justification for racial discrimination, in theory or in practice, anywhere."

Cooperation was the anvil for evolutionary progress, and competition was only a tactic in the history of biological world. Till the humans emerged, it was all about the food chain and feedback loop and, hence, eco-friendly. According to the Bible, man was cursed to earn his food by the sweat of his forehead. Though it is a metaphor, it is true that the human is the only animal that cultivates and rear animals and birds for his food. Man with his intelligence and physical attributes became the most 'successful' animal in the planet, and his numbers increased and inhabited all the continents. Increase of the population and foraging needs were likely reasons for global migration. This also gave way to greater competition. Migration to different parts of the world segregated humans for long periods of time and physical variations due to several environmental factors ensued. These physical variations were construed as race and polygenism for a long time.

As the humans emerged as the dominant species and started ruling the globe, cooperation has given way to competition. Initially, it was the strife between the races, and later between the nations in the name of patriotism, then on grounds of ethnicity. Humans also started killing each other in the name of religion.

Karl Marx was probably right, and even compassionate to religion, when he stated, "Religion is the sign of the oppressed creature, the sentiment of the heartless world and the soul of the soulless condition.

It is the opium of the people." If Marx were living today, he may have amended his statement to: "Religion is the sign of the fanatic. It is the arrogance of the heartless world. It is the dogma of the bolted mind. It is the maddening drug of the people." No more is it the sign of the oppressed creature. As Jonathan Swift said, "We have enough religion to make us hate, but not enough to make us love one another."

Will We Ever Reach Perfection?

Perfection is like a mirage where function, beauty, wisdom and all positive qualities meet and merge. Evolution is an innate quality of life, and may be the quest of nature/matter for perfection, where trial and error, rejection and selection are the norms.

Through the wars between nations, fight between religions, molestation of nature with atomic wastes, suffocating her with pollution and hindering her plans by eugenics, are we on the path of self-inflicted extinction? Are we, as Sir Winston Churchill said, on the verge of "A new dark age made more protracted and perhaps more sinister by the light of perverted science."

Are we misusing the self-awareness and intelligence gained through evolution? If so, we may be subjected to nature's wrath- extinction and a future intelligent species (if any) may rename us 'Homo stupid.'

The history of the biosphere undoubtedly reveals the fact that our 'living planet' rejects the living species that falls short of the requisites or misbehaves, thereby posing a threat to the ecosystem or interrupting the feedback loop. Examples are aplenty. Dinosaurs ruled the world for 135 million years (an impressive period of time) but became extinct around seventy million years ago. The Australopithecus appeared on the planet only four million years ago to remain on the face of the earth for a short span of one million years. Later, Hominids of better intelligence like the Homo habilis, Homo erectus and Homo neanderthalensis emerged for short periods, only to become extinct in less than a million years. All of them belong to our genus. Humans are scientifically known as Homo sapiens meaning the 'wise man'—the so presumed, most evolved Hominid. The modern man (Homo sapiens sapiens) appeared only 250,000 years ago.

Post human

Arthur Clarke's Third Law: "Any sufficiently advanced technology is indistinguishable from magic." *(Arthur C. Clarke, author of* 2001: A Space Odyssey *and many other works of science fiction, was a renowned scientist.)*

Shermer's Last Law: "Any sufficiently advanced extra-terrestrial intelligence is indistinguishable from God." *(Michael Shermer was the founder of the Skeptic Society, and an American science writer and science historian.)*

Biological evolution is very slow and takes very long time for humans to achieve super-human status. However, technological evolution is very fast and that will change our way of life. We took only around 10,000 years to go from wooden wheels to the airplane, and only sixty-six years from flight to landing on the moon. "Ray Kurzweil, in *The Age of Spiritual Machines*, calculates that there have been thirty-two doublings since World

War II. The singularity point may be upon us as early as 2030." (Shermer)

Peter ward, author of *Future Evolution*, is of the opinion that humanity may exist for a long time. He feels that, for a new species to form, humans should isolate and interbreed, which is improbable. There is another possibility—that climatic changes, disease due to the emergence of superbugs, nuclear war and so on could cause the extinction of humans.

Due to scientific progress, and our capabilities to manipulate genes, the emergence of 'designed human beings' with prolonged life and superior intelligence is possible in the near future. Science has already mapped the DNA sequence of the entire human Genome.

The intelligence of the post-human can be either completely artificial or a combination of human and artificial intelligence. It is actually a process of redesigning the human intelligence by nano-technology and genetic engineering, life-extending technologies, psycho-pharmacology and implantable computers.

Kevin Warwick, a leader in cybernetics, envisages the existence of humans and post-humans in the same society. Post-humans will be considered as gods by ordinary men because of their higher intelligence and technological advancements that can overcome the laws of nature.

Julien Huxley coined the word 'transhumanism' in an article in 1957. This could happen by natural evolution or by the creation of intelligent animal species.

Technological changes by genetic engineering, life-extending therapies and so on could beget rapid changes in the human conditions. Transhumanist Philosopher Nick Bostrom warns that this can cause a threat to the existence of humans and other dangers, even existential problems.

Epilogue

A Stroll in the Woods

Occasionally, for a sojourn and to get recharged, I go to the nearby woods. While on the stroll, a cool breeze caressed me and a sweet fragrance came my way. I inhaled deeply. The soothing breeze murmured, "The fragrance belongs to us." I looked around and saw a host of smiling blooms fluttering and dancing in the breeze. They told me, "We gave the fragrance unto the breeze." The twig, which held the flower, said, "She is mine." The tree said, "The branches, twigs and the flowers belong to me." The roots said, "We nourish them all." The ground on which the tree stood said, "I give them water and nutrients—without me how can they be?" The evening sun was smiling through the canopy.

For a moment, I was stunned. To quell my anxiety, they said in unison,

"You stupid man, who call yourself sapiens (wise)? Realize that we all, including you, are parts of this living planet. Take care not to destroy, for your lofty existence as well."

I closed my eyes, listened to the music of the nature and filled my lungs with fragrant fresh air, as the revelation dawned on me that I too am, though insignificant, a part of this vast expanse called 'Infinity.'

Bibliography

A.I. Oparin, *Life, Its Nature, Origin and Evolution*

A.I. Oparin, *The History of the Theory of Genesis and Evolution of Life*

A.I. Oparin, *The Origin of Life on Earth*

Ackermann R.R., Cheverud J.M. Detecting genetic drift versus selection is human evolution. PNAS. 2004; 101:17946-51

All that glitters: A review of physiological research on the aesthetics of the golden section Perception .1995; 24: 937-68

Alles D.L., Stevenson J.C. Teaching human evolution. The American Biology Teacher. 2003; 65(5): 333-339

Alpheus Spring Packard, *Life-History of Animals* (1876)

Alvarez L.W., *Alvarez: Adventures of a Physicist* (1987)

Amanda Spink, *Information Behaviour: An Evolutionary Instinct*, pp. 35–42 (2010)

Arens, N.C. and West, I.D. *Press-pulse: A general theory of mass extinction* (2008)

Arthur Koestler, *The Case of the Midwife Toad* (1971)

Bacon F., Essays, 'Of Beauty' (1625)

Baddeley, A.D., 'The episodic buffer: a new component of working memory?', Trends in Cognitive Science 4 (11): 417–23. doi:10.1016/S1364-6613(00)01538-2. PMID 11058819 (2000)

Behe, Michael J., *Darwin's Black Box: The Biochemical Challenge to Evolution* (1996)

Bill Bryson, *A Short History of Nearly Everything* (2007)

Boussaud, D., Desimone, R., & Ungerleider, L.G., *Visual topography of area TEO in the macaque*. Journal of Comparative Neurology, 306, 554–575. (1991)

Brewer, A.A., Press, W.A., Logothetis, N.K., & Wandell, B.A. *Visual areas in macaque cortex measured using functional magnetic resonance imaging*. Journal of Neuroscience, 22,10416–10426. (2002)

Broadbent B.H. Sr., *A new X-ray technique and its application to orthodontia*. Angle Orthod 1931; 1: 45-66

Bruce L.L., Neary T.J., *The limbic system of tetrapods: a comparative analysis of cortical and amygdalar populations*, Brain Behav. Evol. 46 (4–5): 224–34. doi:10.1159/000113276. PMID 8564465 (1995)

Buchmann, Stephen, *Letters from the Hive: An Intimate History of Bees, Honey, and Humankind*, Random House of Canada, p. 105. ISBN 978-0-553-38266-2 (2006)

Byron C., *A description of fossil hominid and their origins*, Kent state University

Cane M.A., Molnar P., *Closing of the Indonesian seaway as a precursor to East African aridification around 3-4 million years ago*, Nature 2001; 411:157-162

Carmichael, S.T., & Price J.L., *Architectonic subdivision of the orbital and medial prefrontal cortex in the macaque monkey*, Journal of Comparative Neurology, 346, 366–402. (1994)

Charles B. Thaxton, Walter L. Bradley, and Roger L. Olsen, *The Mystery of Life's Origin: Reassessing Current Theories* (Philosophical Library, January 19, 1984)

Charles Darwin, *On the origin of species by means of natural selection* (1859)

Charles Darwin, *The descent of man*—Volumes 1 and 2 (1871)

Charles Darwin, *The expression of the emotions in man and animals* (1872)

Motilal Banarsidass, *Classical Samkhya*, Delhi, 2nd. Ed

Colby, C.L., Gattass, R., Olson, C.R., & Gross, C.G., *Topographic organization of cortical afferents to extrastriate visual area PO in the macaque: a dual tracer study*, Journal of Comparative Neurology, 269, 392–413. (1988)

Cowey, A., Heywood, C.A., & Irving-Bell, L., *The regional cortical basis of achromatopsia: a study on an achromatopsic patient*, European Journal of Neuroscience, 14, 1555–1566 (2001)

Crawford, O.G.S., *Man and his past*, London, Oxford University Press, 1-19 (1921)

Dawkins, Richard, *The God delusion*

Dawkins, Richard, *The selfish gene*

Dawkins, Richard (editor) *The oxford book of modern science writing*

Dawson K.J., *The Advantage of Asexual Reproduction: When is it Two-fold?*, Journal of Theoretical Biology 176 (3): 341–347. doi:10.1006/jtbi.1995.0203

Denys, K., Vanduffel, W., Fize, D., Nelisen, K., Peuskens, H., Van Essen, D., & Orban, G.A., *The processing of visual shape in the cerebral cortex of human and non-human primates: An fMRI study* (2003)

Desimone, R., & Ungerleider, L., *Neural mechanisms of visual processing in monkeys*, In F. Boller & J. Graman (Eds.), *Handbook of Neuropsychology* (pp. 267–299). Amsterdam: Elsevier (1989)

Desmond Morris, *Intimate behavior*

Desmond Morris, *The Naked Ape*

Desmond Morris, *The human zoo*

DeYoe, E. A., Carman, G., Bandetinni, P., Glickman, S., Wieser, J., Cox, R., Miller, D., & Neitz, J., *Mapping striate and extrastriate visual areas in human cerebral cortex*, Proceedings of the National Academy of Sciences, 93, 2382–2386 (1996)

Dietrich, Michael R., *Richard Goldschmidt: hopeful monsters and other heresies*, Nature Reviews Genetics 4 (Jan.): 68-74 (2003)

Dougherty, R.F., Koch, V.M., Brewer, A., Fischer, B., Modersitzki, J., & Wandell, B., *Visual field representations and locations of visual areas V1/2/3 in human visual cortex*, Journal of Vision, 3, 1–3. (2003)

Drury N.E., *Beauty is only skin deep*, Journal of the Royal Society Medicine. 2000; 93: 89-92, Green CD

Eduard Strasburger, *On Cell Formation and Cell Division* (1876)

Edward Drinker Cope, *The Origin of the Fittest: Essays on Evolution* (Nature journal, 1887)

Edward Drinker Cope, *On The Origin of Genera; From the Proceedings of the Academy of Natural Sciences of Philadelphia, Oct. 1868* (Merrihew & Son, 1869)

Edward Willett (October 29, 2008)

Ekman, P; Friesen, W., *Constants across cultures in the face and emotion*, Journal of Personality and Social Psychology 17 (2): 124–9. doi:10.1037/h0030377. PMID 5542557 (1971)

Ernst Haeckel, *Freedom in Science and Teaching* (1879), reprint edition, University Press of the Pacific (2004)

Ernst Haeckel, *The History of Creation* (1868), translated by E. Ray Lankester, Kegan Paul, Trench & Co., London, 1883, 3rd edition, Volume 1

Etcoff N.L., *Psychology: Beauty and the beholder*, Nature 1994; 368:186-87

Etcoff N.L., *Survival of the prettiest: The science of beauty*

Farkas L.G., *Anthropometry of the head and face in medicine*, New York, Elsevier Science Publishing Co. Inc (1981)

Felleman, D.J., & Van Essen, D.C., *Distributed hierarchical processing in primate cerebral cortex*, Cerebral Cortex, 1, 1–47 (1991)

Fellman et al, *Emotional selection and human personality: biological theory*

Ferrario V.F., Sforza C., Poggio C.E., Tartaglia G., *Facial morphometry of television actresses compared with normal women*, J Oral Maxillofac Surg 1995; 53:1008-14

Ferry, A.T., ngür, D., An, X., & Price, J.L., *Prefrontal cortical projections to the striatum in macaque monkeys: evidence for an organization related to prefrontal networks*, Journal of Comparative Neurology, 425, 447–470 (2000)

Fisher, R.A., *Has Mendel's work been rediscovered?*, Annals of Science 1 (2): 115–126 (1936)

Fisher, R.A.; Balmukand, B., *The estimation of linkage from the offspring of selfed heterozygotes*, Journal of Genetics (20): 79–92 (1928)

Fisher, R.A., Corbet, A.S.; Williams, C. B., *The relation between the number of species and the number of individuals in a random sample of an animal population.* Journal of Animal Ecology (12): 42–58 (1943)

Fize, D., Vanduffel, W., Nelissen, K., Denys, K., Chef d'Hotel, C., Faugeras, O., & Orban, G. A., *The retinotopic organization of primate dorsal V4 and surrounding areas: A functional magnetic resonance imaging study in awake monkeys*, Journal of Neuroscience, 23, 7395–7406 (2003)

Francis Crick, *Life Itself: Its Origin and Nature* (Simon & Schuster, 1981)

Freud Sigmund, *A general introduction to psycho analysis*

Freud Sigmund, *Introductory lectures on psycho analysis*

Gardner Murphy, *An introduction to psychology*

Gardenfors Peter, *How homo became sapiens*

Gary Lynch & Richard Grancher, *Big brain; The origins and future of human intelligence*

Gattass, R., Sousa, A. P. B., & Gross, C. G., *Visuotopic organization and extent of V3 and V4 of the macaque*, Journal of Neuroscience, 8, 1831–1845 (1988)

Girard, P., Lomber, S.G., & Bullier, J., *Shape discrimination deficits during reversible deactivation of area V4 in the macaque monkey*. Cerebral Cortex, 12, 1146–1157 (2002)

Grammer K, Thornhill R., *Human (Homo sapiens) facial attractiveness and sexual selection: The role of symmetry and averageness*, Journal of Comparative Psychology, 1994; 108: 233-242

Hoffecker John F., *Landscape of the mind*

Hume David, *A treatise of human nature*

Jared Diamond, *The rise and fall of the third Chimpanzee: How our animal heritage affects the way we live*

Jared Diamond, *Guns, germs and steel, a short history of everybody for the last 13000 years*

Hadjikhani, N., Liu, A. K., Dale, A.M., Cavanagh, P., & Tootell, R.B.H., *Retinotopy and color sensitivity in human visual cortical area V8*, Nature Neuroscience, 1, 235–241 (1998)

Hannah Holmes, *The well dressed Ape*

Harbor Symposia in Quantitative Biology, 55, 679–696

Hawking Stephen W., *A brief history of time: From the big bang to black holes*

Horm Behav, *Conditional expression of women's desires and men's mate guarding across the ovulatory cycle*, 49 (4): 509–18

Hugo de Vries, *The Mutation Theory*

Huk, A.C., Dougherty, R.F., & Heeger, D.J., *Retinotopy and functional subdivision of human areas MT and MST*, Journal of Neuroscience, 22, 7195–7205 (2002)

Huxley, Leonard, *Life and Letters of Sir Joseph Dalton Hooker OM GCSI*. London, Murray (1918)

James Lovelock, *The Revenge of Gaia: Why the Earth is Fighting Back – and How we Can Still Save Humanity* (2006)

J.B.S. Haldane, *Mathematical Darwinism: A discussion of the genetical theory of natural selection*, The Eugenics Review 23 (2): 115–117. (1931)

J.B.S. Haldane, *Science and Human Life*, Harper and Brothers, Ayer Co. reprint (1933)

J.B.S. Haldane, *The Inequality of Man, and Other Essays* (1932)

Jefferson Y., *Skeletal Types. Key to unraveling the mystery of facial beauty and its biologic significance*. J Gen Orthod 1996; 7:7-25

Jerry A. Coyne, *The genetic basis of Haldane's rule*, (PDF) Nature 314 (6013): 736–7388 (25 April, 1985)

Jerry A. Coyne, Faith Versus Fact: Why Science and Religion Are Incompatible. *Viking Press. p. 311 (19 May, 2015)*

Jerry A. Coyne, *Why Evolution is True*, Viking, New York (USA); Oxford University Press, Oxford (UK) (2009-01-22)

John C. Eccles, *Evolution of the brain: Creation of the self*

Jones D.E. *Sexual selection, physical attractiveness and facial neoteny. Cross cultural evidence and implications*, Curr Anthropol 1995; 36: 723-748

Joshua Lederberg and Edward Tatum, (Lederberg J, Tatum EL (1946) *Gene recombination in E. coli*, Nature 158 (4016)

Juvan Mascaro (translated and sected), *The Upanishads* (Penguin classics)

Jung Carl Gustav, *Four Archetypes*

Kagan Jerome, *An argument for mind*

Kanazawa, S., *Intelligence and Physical Attractiveness*. Intelligence, 39 (1), 7-14 (2011)

Kaplan H.S., Robson A.J., *The emergence of humans. The co evolution of intelligence and longevity with intergenerational transfers*, PNAS. 2002; 99: (15): 10221-10226

Kurzwell Ray, *How to create a mind*

Kauffman, S.A., Johnsen, S., *Co-Evolution to the Edge of Chaos: Coupled Fitness Landscapes, Poised States, and Co-Evolutionary Avalanches*, Journal of Theoretical Biology 149: 467–505 (1991)

Kauffman, Stuart, *The Origins of Order: Self Organization and Selection in Evolution*, Oxford University Press (1993)

Kauffman, Stuart, *At Home in the Universe: The Search for Laws of Self-Organization and Complexity*, Oxford University Press (1995)

Keats John, *Book 1 of Endymion*

Kelly, R.M., & Strick, P.L., *Cerebellar loops with motor cortex and prefrontal cortex of a non-human primate*, Journal of Neuroscience, 23, 8432–8444 (2003)

Key R.F., Grine F.E., *Tooth morphology, wear and diet in Australopithecus and Paranthropus from southern Africa. Evolutionary History of the robust Astraulopithecines Grine FE* (Ed), 427-447 (1988)

Klein, Richard G., *Three Distinct Human Populations. Biological and Behavioral Origins of Modern Humans*, Access Excellence @ the National Health Museum, Retrieved 2007-09-10

Klemm W.R., *Mental biology*

Kornberg A, *Enzymatic Synthesis of DNA*, John Wiley & Sons (1961)

Lamarck, J.B., *Zoological Philosophy*, London (1914)

Lamarck, *Zoological Philosophy, an explosion with regard to natural history of animals*

Lancaster, H.O., *Mathematicians in medicine and biology. Genetics before Mendel: Maupertuis and Réaumur*, Journal of medical biography 3 (2): 84–9 (May 1995)

LeDoux, J., *Synaptic Self*, New York: Penguin Books, 0142001783 (2003)

Leiberman D.E., *The evolution of the human head*

Leiberman D.E., Mc Bratney B.M., Krovitz G., *The evolution and development of cranial form in Homo sapiens*, PNAS. 2002; 99: 1134-1139

Leiberman D.E., Ross C.F., Ravosa J.M., *The primate cranial base: Ontogeny, function and integration*, Year Book of Physical Anthropolo. 2000; 43: 117-119

Lewis, J.W., & Van Essen, D.C., *Architectonic parcellation of parieto-occipital cortex and interconnected cortical regions in the Macaque monkey*, Journal of Comparative Neurology, 428, 79–111. (2000)

Luciano Fadiga, Laila Craighero, Maddalena Fabbri Destro, Livio Finos, Nathalie Cotilon-Williams, Andrew T. Smith, and Umberto Castiello, *Language in Shadow.* Social Neuroscience 1(2) (2006)

Lueck, C.J., Zeki, S., Friston, K.J., Deiber, M.P., Cope, P., Cunningham, V.J., Lammertsma, A.A., Kennard, C., & Frackowiak, R.S.J., *The colour centre in the cerebral cortex of man*, Nature, 340, 386–389 (1989)

Lui, J.H., Hansen, D.V., Kriegstein, A.R., *Development and Evolution of the Human Neocortex*, Cell 146(1):1836. doi:10.1016/j.cell.2011.06.030. PMC 3610574. PMID 21729779 (2011)

Lyell, Charles, *Principles of geology, vol. 3. London: John Murray* (1833)

Lyon, D.C., & Kaas, J.H., *Evidence for a modified v3 with dorsal and ventral halves in macaque monkeys*, Neuron, 33, 453–461 (2002)

MacLean P.D., *Psychosomatic disease and the visceral brain; recent developments bearing on the Papez theory of emotion*, Psychosom Med 11 (6): 338–53

Mani. V., *Orthognathic surgery, Esthetic surgery of the face* (Jaypee Brothers medical publishers)

Mani. V., *Surgical correction of facial deformities* (Jaypee brothers medical publishers, 2010)

Margulis, Lynn, *Symbiosis in Cell Evolution: Microbial Communities in the Archean and Proterozoic Eons*, W.H. Freeman (1992)

Margulis, Lynn, *Genome acquisition in horizontal gene transfer: symbiogenesis and macromolecular sequence analysis*, Humana Press. pp. 181–191 (2009)

Margulis, Lynn, and Dorion Sagan, *Acquiring Genomes: A Theory of* (2002); Margulis, Lynn, *Symbiotic Planet : A New Look at Evolution*, Basic Books (1998)

Marzke M.W., Marzke R.F., *Evolution of the human hand: approaches to acquiring, analysing and interpreting the anatomical evidence*, J Anat. 2000 Jul; 197 (Pt1):121-40

Mc Henry H.M., *Tempo and Mode in Human Evolution* Proceedings of the National Academy of Sciences of the United States of America, 1994; 91(15): 6780-6786

McKeefry, D.J., & Zeki, S., *The position and topography of the human colour centre as revealed by functional magnetic resonance imaging*, Brain, 120, 2229–2242 (1997)

McNamara J.A. Jr., *A method of cephalometric evaluation*, Am J Orthod 1984; 86: 449-469

McNeil R.W., Proffit W.R., White R.P., *Cephalometric prediction for orthognathic surgery*, Angle Orthod 1972; 42:154-164

Michael J. Behe, *Irreducible Complexity is an Obstacle to Darwinism Even if Parts of a System Have Other Functions*, Discovery Institute (Feb 18, 2004)

Michel R.W. Dawson, *Mind Body World*

Michael Shermer, *How We Believe & The Borderlands of Science*

Moir, 1989; Tannen, 1991

Napier J.R., *The prehensile movements of the human hand*; J. Bone Joint Surg Br., 1956 Nov; 38-B (4):902-13

ngür, D., Ferry, A.T., & Price, J.L., *Architectonic subdivision of the human orbital and medial prefrontal cortex*, Journal of Comparative Neurology, 460, 425–449 (2003)

ngür, D., & Price, J.L., *The organization of networks within the orbital and medial prefrontal cortex of rats, monkeys and humans*, Cerebral Cortex, 10, 206–219 (2000)

Nietzsche Friedrich, *Selected Writings*

Noback, Strominger, Demarest, Ruggiero, Charles, Norman, Robert, David, *The Human Nervous System: Structure and Function*. Totowa, NJ: Humana Press. p.25. ISBN 1-59259-730-0 (2005)

Perlman, David, *Fossils From Ethiopia May Be Earliest Human Ancestor*, National Geographic News (July 12, 2001)

Perrett, D.I., Lee, K.J., Penton-Voak, I.S., Rowland, D.R., Yoshikawa, S., Burt, D.M., Henzi, S.P., Castles, D.L., Akamatsu, S. et al., *Effects of sexual dimorphism on facial attractiveness*. Nature 394 (6696): 884–7. doi:10.1038/29772. PMID 9732869 (1998)

Petrides, M., & Pandya, D.N., *Comparative cytoarchitectonic analysis of the human and the macaque ventrolateral prefrontal cortex and corticocortical connection patterns in the monkey*. European Journal of Neuroscience, 16, 291–310 (2002)

Philip Lieberman, *On the Nature and Evolution of the Neural Bases of Human Language*, Yearbook of Physical Anthropology 45: 36–62. PMID 12653308 (2002)

The Language Instinct, Penguin, England (1994)

Pinker, S., *How the mind works*

Piyasīlo, *The Buddha's teachings: a study of comparative Buddhism in truth, tradition & transformation*, Dharmafarer integrated syllabus series (4) (2 ed.), Dharmafarer Enterprises, p. 130 (1991)

Poirier F.E., Mc Kee J.K., *Understanding Human Evolution* Upper Saddle River, NJ 4th Ed (1999)

Potts R., *Environmental hypothesis of Hominin evolution*. Year book of Physical Anthropolo. 41: 91-136 (1988)

Press, W.A., Brewer, A.A., Dougherty, R.F., Wade A.R., & Wandell, B.A., *Visual areas and spatial summation in human visual cortex*, Vision Research, 41, 1321–1332 (2001)

Ramachandran V.S., *The tell-tale Brain: Unlocking the mystery of human nature*

Ramachandran V.S., *Phantoms in the brain*

Ramachandran V.S., *A brief tour of human consciousness*

Raup, David, *Extinction: Bad Genes or Bad Luck?* (1992)

Raup, David, *The Nemesis Affair: A Story of the Death of Dinosaurs and the Ways of Science* (1999)

Raup, David M. and Sepkoski, J. John, Jr., *Mass extinctions in the marine fossil record*, Science 215 (4539) (19 March, 1982)

Ricketts R.M., *The biologic significance of the divine proportion and Fibonacci series*, Am.J.Orthod 81:351-7 (1982)

Ridley Matt, *The Red queen, Sex and the evolution of human nature*

Rightmire G.P., *The evolution of Homo erectus*, Cambridge University press, Cambridge, UK (1990)

Rikowski A., Grammer K., *Human body odour, symmetry and attractiveness*, Proceedings of the Royal Society B 266 (1422): 869–74 (May 1999) and Haselton M.G., Gangestad S.W. (April 2006)

Robert Shapiro, *Planetary Dreams: The Quest to Discover Life Beyond Earth*, Wiley, 1st edition (May 18, 2001)

Robert Shapiro, *Origins: A Skeptic's Guide to the Creation of Life on Earth*, Summit Books (January 1986)

Rock W.P., Sabicha A.M., Evas Riw, *A cephalometric comparison of skull from and the fourteenth, sixteenth and twentieth centuries*, Brit Dent J.; 1: (2006)

Roger Penrose, *Cycles of time: What came before Big Bang*

Rothenberg, *Survival of the beautiful: Art, Science and Evolution*

Roth Gerhard, *The long evolution of Brains and Minds*

Russon A.E., Begun D.R., (editors) *The Evolution of thought, Evolution of great ape intelligence*

Sagan Carl, *Dragons of Eden*

Sagan Carl, *Brocas brain; Reflections on the romance of science*

Sagan Carl, *Cosmos*

Sagan Carl, *Shadows of forgotten ancestors*

Sandler, I., *Pierre Louis Moreau de Maupertuis - a precursor of Mendel*, Journal of the history of biology 16 (1): 101–36 (1983)

Sarich V.M., Wilson A.C., *Immunological time scale for hominid evolution*, Science. 158:1200-1203 (1967)

Schacter, Daniel L., Psychology, sec. 3.20 (2012)

Schrodinger Erwin, *What is life, with Mind and matter and Autobiographical sketches*

Sepkoski, J. John, Jr., *A kinetic model of Phanerozoic taxonomic diversity. III. Post-Paleozoic families and mass extinctions*, Paleobiology 10 (2): 246–267 (1984)

Sereno, M.I., Dale, A.M., Reppas, J.B., Kwong, K.K., Belliveau, J.W., Brady, T.J., et al., *Borders of multiple visual areas in humans revealed by functional magnetic resonance imaging.* Science, 268, 889–893 (1995)

Seyfarth, Robert M., Cheney, Dorothy L., Marler, Peter, *Vervet monkey alarm calls: Semantic communication in a free-ranging primate*, Animal Behaviour 28 (4): 1070–1094 (1980)

Shakespeare, *Merchant of Venice*

Sidney W. Fox, *Introduction to protein chemistry*, New York: Wiley (1957)

Sidney W. Fox, *The emergence of life: Darwinian evolution from the inside*, Basic Books (1988)

Singh D., *Adaptive significance of female physical attractiveness: role of waste to hip ratio J Personality and social Psycology*, 293-307 (1993)

Smith G.E., *Essays on evolution of man*, pp 145

Smith, A.T., Greenlee, M.W., Singh, K.D., Kraemer, F.M., & Hennig, J., *The processing of first- and second-order motion in human visual cortex assessed by functional magnetic resonance imaging (fMRI)*, Journal of Neuroscience, 18, 3816–3830 (1998)

Smith, Jos A., Cheryl Bardoe, Smith, Joseph A., *Gregor Mendel: The friar who grew peas*, Abrams Books for Young Readers (2006)

Spencer, Herbert, *Education: Intellectual, Moral, and Physical* (1891)

Spencer, Herbert, *Social Statics, Abridged and Revised: Together with the Man Versus the State* (1896)

Spencer, Herbert, *The Study of Sociology*

Spoor C.F., O'higgins P., Dean M.C., Lieberman D.E., *Anterior sphenoid of modern humans*, Nature. 397- 572 (1999)

Sri Aurobindo, *The Life Divine*, Sri Aurobindo Ashram Trust, (1977)

Steiner C.C., *The use of cephalometrics as an aid to planning and assessing orthodontic treatment*, Am J Orthod 1960; 46: 721-735

Suzana Herculano, *The human advantage, A new understanding of how our brain became remarkable*

Thomas H. Morgan, *Evolution and Adaptation*, New York: Macmillan (1903)

Thomas Robert Malthus, *An Essay on the Principle of Population* (1798)

Thornhill, R. Science 205, 412-414 (1979)

Thornhill R., Gangestad S.W., *The scent of symmetry: A human sex pheromone that signals fitness?* J Evolution and Human Behavior, 20(3)175-201 (1999)

Tobias P.V., Olduvai Gorge, *The skulls, endocasts and teeth of H.habilis*, Cambridge University Press, UK

Tobias P.V., Olduvai Gorge, *The cranium and maxillary dentition of Australopithecus boisei*, Cambridge University Press Cambridge, UK (1967)

Tootell, R.B.H., & Hadjikhani, N. *Where is "dorsal V4" in human visual cortex? Retinotopic, topographic and functional evidence*. Cerebral Cortex, 11, 298–311 (2001)

Tovee M.J., Reinhardt S., Emery J.L., Cornelissen P.L. *Optimum body-mass index and maximum sexual attractiveness*, Lancet; 352(9127):548 (Aug 15, 1998)

Tsao, D.Y., Vanduffel, W., Sasaki, Y., Fize, D., Knutsen, T.A., Mandeville, J. B., et al., *Stereopsis activates V3A and caudal intraparietal areas in macaques and humans*, Neuron, 39, 555–568 (2003)

Turrill W.B., *Pioneer Plant Geography: The Phytogeographical Researches of Sir Joseph Dalton Hooker* (1953)

Tweed, Thomas A., American encounter with Buddhism, 1844-1912: Victorian culture & the limits of dissent, UNC Press Books, pp. 107–108. ISBN 978-0807849064 (2000)

Van Essen, D.C., *Surfacebased atlases of cerebellar cortex in human, macaque, and mouse*. Annals of the New York Academy of Sciences, 978, 468–479 (2002)

Van Essen, D.C., *Organization of visual areas in macaque and human cerebral cortex*. In L. Chalupa & J.S. Werner (Eds.), The Visual Neurosciences (pp. 507–521), Cambridge, MA: MIT Press (2004)

Van Essen, D.C., & Zeki, S.M., *The topographic organization of rhesus monkey prestriate cortex*, Journal of Physiology, 277, 193–226 (1978)

Van Essen, D.C., Dickson, J., Harwell, J., Hanlon, D., Anderson, C.H., & Drury, H.A., *An integrated software system for surface-based analyses of cerebral cortex*, Journal of American Medical Informatics Association, 8, 443–459 (2001)

Van Essen, D.C., Felleman, D.F., DeYoe, E.A., Olavarria, J.F., & Knierim, J.J., *Modular and hierarchical organization of extrastriate visual cortex in the macaque monkey*. Cold Spring Harbor Symposia in Quantitative Biology, 55, 679–696 (1990)

Van Essen, D.C., Harwell, J., Hanlon, D., & Dickson, J. *Surface-based atlases and a database of cortical structure and function*, In S. H. Koslow & S. Subramaniam (Eds.), *Databasing the Brain: From Data to Knowledge (Neuroinformatics)*, New York: John Wiley & Sons, NJ (2004)

Van Essen, D. C., Maunsell, J. H. R., & Bixby, J. L., *The middle temporal visual area in the macaque: Myeloarchitecture, connections, functional properties and topographic organization*, Journal of Comparative Neurology, 199, 293–326 (1981)

Van Essen, D.C., Newsome, W.T.N., & Bixby, J. L., *The pattern of interhemispheric connections and its relationship to extrastriate visual areas in the macaque monkey*, Journal of Neuroscience, 2, 265–283 (1982)

Van Essen, D.C., Newsome, W.T., Maunsell, J.H.R., & Bixby, J. L., *The projections from striate cortex (V1) to areas V2 and V3 in the macaque monkey: Asymmetries, areal boundaries, and patchy connections*, Journal of Comparative Neurology, 244, 451–480 (1986)

Van Oostende, S., Sunaert, S., Van Hecke, P., Marchal, G., & Orban, G. A., *The kinetic occipital (KO) region in man: An fMRI study*, Cerebral Cortex, 7, 690–701 (1997)

Verma, O.P.S, Agarwal O.V.K., *Cell Biology, Genetics, Evolution and Ecology*

Wade, A.R., Brewer, A.A., Rieger, J.W., & Wandell, B.A.. *Functional measurements of human ventral occipital cortex: Retinotopy and color*, Philosophical Transactions of the Royal Society, 357, 963–973 (2002)

Walker, A.E., *A cytoarchitectural study of the prefrontal areas of the macaque monkey*, Journal of Comparative Neurology, 73, 59–86 (1940)

Wallace, Alfred Russel, *Darwinism: An Exposition of the Theory of Natural Selection, with Some of Its Applications*, Macmillan (1889)

Wallace, Alfred Russel, *Man's place in the universe* (Gutenberg) (1904)

Wallace, Alfred Russel, *World of Life*, The Alfred Russel Wallace Page hosted by Western Kentucky University

Watson James D., *The Annotated and Illustrated Double Helix*, edited by Alexander Gann and Jan Witkowski (2012)

Watson, J.D., Gunther S. Stent, ed. *The Double Helix: A Personal Account of the Discovery of the Structure of DNA*, W.W. Norton & Company (1968)

Watson, J.D.G., Myers, R., Frackowiak, R.S.J., Hajnal, J.V., Woods, R. P., Mazziotta, J. C., et al.. *Area V5 of the human brain: Evidence from a combined study using positron emission tomography and magnetic resonance imaging*, Cerebral Cortex, 3, 37–94 (1993)

White T.D., Suwa G., Asfaw B.. *Australopithecus ramidus, a new species of early hominid from Aramis, Ethiopia*, Nature, 371(6495):306–312 (22 September, 1994)

William Baetson, *Materials for the study of variation*

William Bateson, *Mendel's principles of heredity*

Williams, Chris, *Stem cell fraudster made 'virgin birth' breakthrough: Silver lining for Korean science scandal*, The Register, (3 August, 2007)

Wood B, Richard B.G., *Human Evolution. Taxonomy and Paleobiology*. J Anatomy; 196: 19-60 (2000)

Index

A.I. Oparin. 12, 18
A. Kornberg. 124
Action Potential. 20
Adenosine triphosphate (ATP). 15
Advaitha. 4, 79
Aerobic metabolism. 205
Alfred Russel Wallace. 81, 106, 112, 254
Allopatric speciation. 145, 150
Alpheus Spring Packard. 110
Alvarez L.W. 120
Amino acids. 11, 89, 101, 123
Amygdala. 30, 39, 63, 78
Anaerobic metabolism. 205
Antigen. 101, 102
Antiserum. 102
Aphasia. 52
Aposematism. 172
Archaeopteryx lithografica. 95

Ardipithecus. 131 134,135.
Arens N.C. 237
Arthanareeswaran. xxxv
Arthur Koestler. 111
Asexual reproduction. 22, 129, 163
Atman. Xxxv, 71,76
Australopithecus afarensis. 90, 131, 136, 137, 156, 161, 195, 200
Australopithecus africanus. 131, 138, 139, 156, 210, 217
Awareness. 45, 67, 74, 83
Axons. 20, 27, 29
B. Bryson. 103
Bacon F. 205
Basal ganglia. 30, 37
Bateson. 121
Beauty. xxix, xxx, 163-182
Beckoning breast. 177
Behe, Michael. 6, 7
Big Bang. Xxxii, xxxiii
Biochemistry, 101
Biopoesis. 13

Biosphere. 82, 86, 230
Biotechnology. xv
Bishop of Worcester. 107
Brahma. 3, 5, 71
Brain stem. 37, 65, 66, 79
Broca's area. 41, 54-56, 145
Buddhist philosophy. 52
Building blocks of life. Xxxiii, 9, 13
C. Lyell, 108
Captain Robert Fitzroy. 106
Carl Correns. 122, 126
Carl von Linne. 96
Carol W. Greiderd. 126
Catalysts. 10
Cephalometric analysis. 193
Cerebellum. 31, 34, 37
Cerebral cortex. 28, 31, 32, 40
Cerebral hemispheres. 29, 31
Charles Darwin 11, 12, 52, 54, 73, 77, 89, 96, 103, 106-114
Chastity. 170, 171
Chemotaxis. 20
Chloroplast, 17, 18, 115
Chromosomes. 101-103, 121, 123, 126
Cingulate cortex. 39.
Clarke. 231
Coacervate. 13
Comparative anatomy. 96
Comparitive embryology 98, 99
Complexing force. 83, 105
Conscience. 46, 65
Conscious mind. 47
Consciousness. 28, 42, 45, 46, 65–87
Convergent evolution. 98
Correns. 122, 126
Cro-magnon man. 154, 155
Darwin C. 11, 12, 52, 54, 73, 77, 89, 96, 103, 106-114
Dating of fossils. 92
David Ferrier. 32
David M. Raup. 93
Dawkins R. 7, 9, 67, 114, 228
Denisovans. 148, 150
Diencephalon. 37
Dinosaurs. xxv, xxxiv, 37,

86, 91, 94, 111, 118, 119, 120, 147, 230

Distraction displays. 55

Divine proportion. xxix, xxx, 173, 208, 214, 215, 218, 219

DNA. xxiii, 1, 7, 14, 16, 101, 102, 123, 126, 132, 134, 165, 225

DNA sequencing. 125, 126

Donald Johanson. 136

Dreams. 38, 48

Dualism. 4, 69

Dubois. 161

Dwaitha. 4

Echolocation. 31

Edison. 226

Edward Drinker Cope. 110

Edward Hitzig. 32

Ego. 42, 45, 46, 47

Elizabeth H Blackburn. 126

Embryology. 92, 98

Endo symbiotic theory.115

Endosymbiosis. 17, 18

Epiphenomenalism, 69

Eran Meshorer. 153

Ernst Haeckel, 99, 110

Esteem. 50, 51

Etcoff Nancy. 204

Ethnologists. 54

Ethology. 73

Etienne Geoffrey Saint-Hilaire. 98

Eugenics. xvii, 87, 230

Evolution 12, 13, 19, 25, 58, 66, 67, 76, 80, 81, 86, 89-129

Evolution of brain. 77, 172

Evolution of Consciousness 66

Evolution of Emotions. 77

Evolution of Language. 58

Evolution of spirituality. 80

Extinction. 91, 118-120

Facial Changes. 159, 193-203

Fixed Action Pattern. 42, 54

Foramen Magnum. 139, 141, 159

Forebrain. 29

Forkhead box protein. 58

Fossil records. 92- 95

Fossilization. 92

Francesco Redi. 10

Francis Crick. 12, 123

Frankort horizontal line.

194
Frederic Sanger. 125
Freudian slip. 48
G.B. Beadle. 124
Gaia hypothesis. 85, 115
Gamete. 76, 114. 123, 165
Gene. 101, 102, 120
Genetics. 120-129
Genome. 2, 126, 127, 154, 227
Geographical Distribution. 96, 107
George Bernard Shaw. 111
Germanism. 228
Golden proportion. 173, 178, 214, 218
Goldilocks Planet. 8
Great oxygenation event. 116
Gregor Mendel. 113, 121, 122, 126
Gustav Fritsch. 32
H.J. Muller. 124
H. floresiensis. 146, 147
Haekelian concept. 100,
Haldane J.B.S. 12, 13, 18, 114
Hard problem. 69, 70 71
Henry, A.G. 153

Herbert Boyer. 126
Herbert Spencer 110, 111
Hierarchy of Needs. 48-52
Hindbrain. 31
Hippocampus 30, 35, 39, 62, 63
HMS Beagle. 106, 107
Homo sapiens idaltu. 134, 151. 152
Homeostasis. 19, 86
Hominidae. 117, 133, 159
Homo antecessor. 134, 147
Homo erectus. 143
Homo ergaster. 142
Homo floresiensis 146
Homo habilis. 56,131, 140
Homo heidelbergensis. 148, 149, 156
Homo neanderthalensis. 150, 152, 156, 231
Homo rudolfensis. 142
Homo sapiens. 151, 156, 169,200,212,213
Homologous structures. 96, 97
Hopeful monster. 129
Hughling Jackson. 32
Hugo de Vries. 121, 126, 127

Index

Humming birds. 34
Huxley Thomas. 70, 73, 84
Hypothalamus. 30, 39
Iceberg model of mind. 46, 47
Id. 42-44
Incarnations. xxxvi
Indian philosophy. 51, 78
Individuation. 82
Inner want. 83, 105
Instinct. 42-44, 77
Insular dwarfing. 146, 147
Intelligence. xxxv, 34, 66
Intuition. 1, 79, 80
J. Lederbug. 124
Jack Sepkoski. 93
Jack W. Szostak. 126
James D. Watson. 123
James Lovelock. 85, 115
Jean-Baptiste Lamarck. 83, 84, 104-110
Jerry Coyne. 8
Joseph Dalton Hooker. 11, 108
Judo-Christian. 3
Kali Yuga. Xxvii, 5
Kalpa. 4, 5
Keats John. 204
Kenyanthropus platyops. 131, 137, 138
Khorana H. 124
Kipling. 228
KT extinction. 5, 119
Lamarck.83, 84, 104-110
Language. 52-61
Larynx. 57-59
Limbic Brain. 35, 38-40
Lips. 179, 201,202
Liran Carmel. 153
Louis Pasteur. 10
Lower third of the Face. 188-194
Lucy. 87, 90
Lynn Margulis. 17, 85, 115
Major Extinction Events. 118-120
Malthus. 107, 108
Mammalian brain. 35, 40
Manvantaras. 4.
Marriage. 170
Maupertuis. 104
Meave Leaky. 142
Mechanism of evolution. 19, 20
Meiosis. 123,127
Memory. 61-64
Mendel Gregor. 113, 121, 122, 126

Meteorite. 11, 12
Midbrain. 31
Middle third of the face. 187
Milky Way. xxxii, 9
Miller-Urey experiment. 13
Mitochondria. 17, 18, 134
Molecular clock. 132
Monism. 4, 69
Morality. 45, 81
Murchinson meteorite. 11
Mutation. 1, 85, 114, 120, 127-129
Neo Darwinism. 126-129
Neo-cortex. xxxv, 34, 35, 40, 41
Neo-Darwinian. 2
Neotenous character. 146
Neoteny. 208
Nerve cords. 23, 24
Nerve net. 20-22
Nerve ring. 22, 23
Nervous system. 19, 28
Neuronal correlates of consciousness. 72
Nirenberg. 124
Nociceptors. 24
Nose. 202
Nucleic acids. 2, 14

Nucleus accumbens. 39
Ontogeny. 100
Opsin. 114
Orbit and eye. 201
Organic evolution. 89, 91
Oriental languages. 58
Orrorin. 131, 133
Orthodontics. 193
Orthognathic surgery. 193
Ocelli. 24
Paleo-anthropology. 132, 157
Pangenesis. 110, 111
Panpsychism. 70
Panspermia. 11, 12
Parthenogenesis. 163, 164
Paul Broca. 32, 52, 198
Paul MacLean. 35, 36
Pelvis. 160, 177
Perfection. xxxv, 82, 83, 230
Phi proportion. 173
Photosynthesis. 91, 116
Phyletic evolution. 128
Phylogeny. 100
Physicalism. 69
Physiological needs. 49, 51
Pineal gland. 30
Pleasure principle. 44, 165

Pluri-potential cells. xvii
Polyploidy. 129
Ponginae. 133
Post human. 231
Preconscious mind. 47
Primates. 29, 35, 88
Primordial soup. 12, 13
R.A. Fischer. 114
Racial, ethnic and religious. 226
Rapid eye movement. 38
Raup D. 120
Rebirth. 4, 76, 83
Recapitulation. 100
Regulatory genes. 128
Reptilian brain. 35, 37, 38
Reptin. 154
Ribosomes. 14, 15
Richard Dawkins. 7, 9, 67, 114, 228
Richard Goldschmidt. 129
Richard Leaky. 142
Richter. 11
RNA. xxxiii, 14, 16, 18
Robert Shapiro. 12
Russell Wallace. 103, 173
S. Ochoa. 124
Safety needs. 50
Sagan Carl. 206
Saharan pump theory. 145
Self-Actualization. 50–52
Sepkoski J.J. 120
Sexual reproduction. 2, 90, 116, 163, 165, 166, 207
Sexual selection. xxx, 111, 112, 166, 168
Shakespeare. 204
Shaw. 111
Shermer. 231, 232
Shynia Yamanaka. 126
Sidney W. Fox 14, 251
Singularity. xxxii
Sir John B. Gurdon. 126
Slave traders. 228
Sleep. 38, 62
Social science model. 48
Soft inheritance. 105
Somatic cells. 105
Soul. 4, 71
Specified complexity. 6, 7
Speech. 52-58
Spencer Herbert. 110, 111
Spontaneous generation. 10, 11
Stuart Kauffman. 15
Steatopygia. xxxi
Stem cells. 22, 164
Strausberger. 124

String theory. xxxiii
Stroll in the woods. 235
Structural model of mind. 41-46
Subconscious mind. 45, 207
Super ego. 45,46
Super mind. 82
Supra-laryngeal vocal tract. 56
Sutton. 124
Symbiogenesis. 86
T.H. Morgan. 124
Taxonomy. 96
Temple of Nature. 104
Thalamus. 30, 78
Thomas Robert Malthus. 107,108
Thumb. 158,159,223
Time line of evolution. 117
Topographical model of mind. 46-48
Transhumanism. 232
Transitional forms. 95
Triune brain. 35,36
Unconscious. 38, 47,48
Upper third of the face. 187
Vedanta. 71
Venus. 174, 183, 224
Vestigial organs. 8, 98
Vitruvian Man. 181, 182
Vocal calls. 84
Von Baer. 100
Vries Hugo. 121, 126, 243
Wallace Russel. 81,103, 106, 112,173, 254
Walter Gilbert. 125
Watson D James. 123,124
Watson John. 48
Wernick's area. 56, 59
West I.D. 94, 237
William Bateson. 121, 255
Worcester (bishop). 107
World war. 49, 226
Yin and Yang. 6
Zionism. 228

www.ingramcontent.com/pod-product-compliance
Lightning Source LLC
Chambersburg PA
CBHW020629220526
45464CB00001B/80